A. S. Eddington and the unity of knowledge:
scientist, Quaker & philosopher

Sir Arthur Stanley Eddington (1882–1944) was a key figure in the development of modern astrophysics, who also made important contributions to the philosophy of science and popular science writing. The Arthur Eddington Memorial Trust was set up after his death in order to hold annual lectures on the relationship between scientific thought and aspects of philosophy, religion or ethics. This 2013 collection gathers together six of these lectures, together with Eddington's 1929 Swarthmore Lecture on *Science and the Unseen World.* A preface written by the Astronomer Royal, Lord Martin Rees, is also included. This is a fascinating book that will be of value to anyone with an interest in the philosophy of science and Eddington's legacy.

VOLKER HEINE was born in 1930 in Hamburg, Germany but moved to New Zealand at age 9 where he grew up mostly under the care of the Veitch family. He gained BSc and MSc degrees from the University of Otago and in 1954 came to Cambridge to study for a PhD under Professor Sir Nevill Mott. In 1955 he married Daphne and their three children were born in Cambridge.

In Cambridge Volker helped to develop what is now an eminent and substantial research group on Theory of Condensed Matter. His research focused on the behaviour of various kinds of material at the basic atomic and quantum level, particularly metals, alloys, silicon, magnets, glass and rocks, with an emphasis on computer simulation and calculation to relate the theory to actual observation. He became a Fellow of Clare College in 1960, and was elected a Fellow of the Royal Society in 1974. As an undergraduate Volker joined the Quakers and he remains involved with them today. He is Chairman of the Arthur Stanley Eddington Memorial Trust under whose auspices this new volume is being published.

A. S. Eddington
and the unity of knowledge:
scientist, Quaker & philosopher

A Selection of the
Eddington Memorial Lectures

EDITED BY
Volker Heine FRS

PREFACE BY
Lord Martin Rees FRS
Astronomer Royal

CAMBRIDGE
UNIVERSITY PRESS

University Printing House, Cambridge CB2 8BS, United Kingdom

One Liberty Plaza, 20th Floor, New York, NY 10006, USA

477 Williamstown Road, Port Melbourne, VIC 3207, Australia

314-321, 3rd Floor, Plot 3, Splendor Forum, Jasola District Centre, New Delhi - 110025, India

103 Penang Road, #05-06/07, Visioncrest Commercial, Singapore 238467

Cambridge University Press is part of the University of Cambridge.

It furthers the University's mission by disseminating knowledge in the pursuit of education, learning and research at the highest international levels of excellence.

www.cambridge.org
Information on this title: www.cambridge.org/9781107037373

First published 2013

A catalogue record for this publication is available from the British Library

ISBN 978-1-107-03737-3 Hardback
ISBN 978-1-107-69212-1 Paperback

Preface by Martin Rees

EMERITUS PLUMIAN PROFESSOR

ARTHUR STANLEY EDDINGTON (it seems that he used both forenames in different communities) (1882 – 1944) was Plumian Professor of Astronomy and Experimental Philosophy, and Fellow of Trinity College, at Cambridge University. He was a versatile theorist and one of the fathers of modern astrophysics, best remembered for his key ideas on the nature of stars. He led the way in propounding Einstein's theory of relativity to the English speaking world, and directed the famous expeditions to test this theory by observing stars during the 1919 solar eclipse. His highly successful popular books brought these findings to the attention of a general readership. In a fine memoir entitled *The Greatest Astrophysicist of his Time* (published by CUP) S. Chandrasekhar gives a fuller assessment of his scientific achievements and influence.

The present volume relates to another side of Eddington. He was a lifelong Quaker, and the resonances between scientific, philosophical and religious ways of seeking truth were very important to him. Three series of lectures that he gave on this theme have recently been reprinted by Cambridge University Press: the Gifford Lectures *The Nature of the Physical World* 1928, the Messenger Lectures *New Pathways in Science* 1935, and the Tarner Lectures *The Philosophy of Physical Science* 1939.

Eddington saw early on, and explained very clearly, the philosophical implication of relativity and the new quantum mechanics of the 1920s for the conceptual basis of physics and of science more widely, and for his philosophical beliefs. The connections between these strands of his thinking are eloquently expounded in his 1929 Swarthmore Lecture to the Society of Friends (Quakers) in Britain, which is the first item included in the present volume.

After his death in 1944, some money was collected and the Arthur Stanley Eddington Memorial Trust was set up in order to hold a series of annual lectures on:

> "some aspect of contemporary scientific thought considered in its bearing on the philosophy of religion or on ethics. It is hoped that they will thus help to maintain and further Eddington's concern for relating the scientific, the philosophical and the religious methods of seeking truth and will be a means of developing that insight into the unity underlying these different methods which was his characteristic aim."

The most influential of these Memorial Lectures was *An Empiricist's View of Religious Belief* given by the philosopher Richard Brathwaite. This lecture is reprinted in the present volume, along with five others. For various reasons, the series became non-viable after the 1980s. However Eddington continues to be commemorated by a special annual lecture on astronomy at the Cambridge Institute of Astronomy, the successor institution to the Cambridge Observatory which he directed for 30 years.

SCIENCE AND THE UNSEEN WORLD

❋❋
❋

A. S. Eddington

FRS

THE TWENTY-SECOND

SWARTHMORE LECTURE

22 July 1929

Synopsis

Science and the Unseen World

LOOKING BACK THROUGH THE long past we picture the beginning of the world – a primeval chaos which time has fashioned into the universe that we know. Its vastness appals the mind; space boundless though not infinite, according to the strange doctrine of science. The world was without form and almost void. But at the earliest stage we can contemplate the void is sparsely broken by tiny electric particles, the germs of the things that are to be; positive and negative they wander aimlessly in solitude, rarely coming near enough to seek or shun one another. They range everywhere so that all space is filled, and yet so empty that in comparison the most highly exhausted vacuum on earth is a jostling throng. In the beginning was vastness, solitude and the deepest night. Darkness was upon the face of the deep, for as yet there was no light.

The years rolled by, million after million. Slight aggregations occurring casually in one place and another drew to themselves more and more particles. They warred for sovereignty, won and lost their spoil, until the matter was collected round centres of condensation leaving vast empty spaces from which it had ebbed away. Thus gravitation slowly parted the primeval chaos. These first divisions were not the stars but what we should call "island universes", each ultimately to be a system of some thousands of millions of stars. From our own island universe we can discern the other islands as spiral nebulae lying one beyond another as far as the telescope can fathom. The nearest of them is such that light takes 900,000 years to cross the gulf between us. They acquired rotation (we do not yet, how) which bulged them into flattened form and made them wreathe themselves in spirals. Their forms, diverse yet with underlying regularity, make a fascinating spectacle for telescopic study.

As it had divided the original chaos, so gravitation subdivided the island universes. First the star clusters, then the stars themselves were separated. And with the stars came light, born of the fiercer turmoil which ensued when the electrical particles were drawn from their solitude into dense throngs. A star is not just a lump of matter casually thrown together in the general confusion; it is of nicely graded size. There is relatively not much more diversity in the masses of new-born stars than in the masses of new-born babies. Aggregations rather greater than our Sun have a strong tendency to sub-divide, but when the mass is reduced a little the danger quickly passes and the impulse to sub-division is satisfied. Here it would seem the work of creation might cease. Having carved chaos into stars, the first evolutionary impulse has reached its goal. For many billions of years the stars may continue to shed their light and heat through the world, feeding on their own matter which disappears bit by bit into ætherial waves.

Not infrequently a star, spinning too fast or strained by the radiant heat imprisoned within it, may divide into two nearly equal stars, which remain yoked together as a double star; apart from this no regular plan of further development is known. For what might be called the second day of creation we turn from the general rule to the exceptions. Amid so many myriads there will be a few which by some rare accident have a fate unlike the rest. In the vast expanse of the heavens the traffic is so thin that a star may reasonably count on travelling for the whole of its long life without serious risk of collision. The risk is negligible for any individual star; but ten thousand million stars in our own system and more in the systems beyond afford a wide playground for chance. If the risk is one in a hundred millions some unlucky victims are doomed to play the role of "one". This rare accident must have happened to our Sun – an accident to the Sun, but to us the cause of our being here. A star journeying through space casually overtook the Sun, not indeed colliding with it, but approaching so close as to raise a great tidal wave. By this disturbance jets of matter spurted out of the Sun; being carried round by their angular momentum they did not fall back again but condensed into small globes – the planets. By this and similar events there appeared here and there in the universe something outside Nature's regular plan, namely a lump of matter small enough and dense enough to be cool. A temperature of ten million degrees or more prevails through the greater part of the interior of a star; it cannot be otherwise so long as matter remains heaped in immense masses. Thus the design of the first stage of evolution seems to have been that matter should ordinarily be endowed with intense heat. Cool matter appears as an afterthought. It is unlikely that the Sun is the only one of the starry host to possess a system of planets, but it is believed that such development is very rare. In these exceptional formations Nature has tried the experiment of finding what strange effects may ensue if matter is released from its usual temperature of millions of degrees and permitted to be cool.

Out of the electric charges dispersed in the primitive chaos ninety-two different kinds of matter – ninety-two chemical elements – have been built. This building is also a work of evolution, but little or nothing is known as to its history. In the matter which we handle daily we find the original bricks fitted together and cannot but infer that somewhere and somewhen a process of matter-building has occurred. At high temperature this diversity of matter remains as it were latent; little of consequence results from it. But in the cool experimental stations of the universe the differences assert themselves. At root the diversity of the ninety-two elements reflects the diversity of the integers from one to ninety-two; because the chemical characteristics of element No. 11 (sodium) arise from the fact that it has the power at low temperatures of gathering round it eleven negative electric particles; those of No. 12 (magnesium) from its power of gathering twelve particles; and so on.

It is tempting to linger over the development out of this fundamental beginning of the wonders studied in chemistry and physics, but we must hurry on. The provision of certain cool planetary globes was the second impulse of evolution, and it has exhausted itself in the formation of inorganic rocks and ores and other materials. We must look to a new exception or abnormality if anything further is to be achieved. We can scarcely call it an accident that among the integers there should happen to be the number 6; but I do not know how otherwise to express the fact that organic life would not have begun if Nature's arithmetic had overlooked the number 6. The general plan of ninety-two elements, each embodying in its structural pattern one of the first ninety-two numbers, contemplates a material world of considerable but limited diversity; but the element carbon, embodying the number 6, and because of the peculiarity of the number 6, rebels against limits. The carbon atoms love to string themselves in long chains such as those which give toughness to a soap-film. Whilst other atoms organise themselves in twos and threes or it may be in tens, carbon atoms organise themselves in hundreds and thousands. From this potentiality of carbon to form more and more elaborate structure a third impulse of evolution arises.

I cannot profess to say whether anything more than this prolific structure-building power of carbon is involved in the beginning of life. The story of evolution here passes into the domain of the biological sciences for which I cannot speak, and I am not ready to take sides in the controversy between the Mechanists and the Vitalists. So far as the earth is concerned the history of development of living forms extending over nearly a thousand million years is recorded (though with many breaks) in fossil remains. Looking back over the geological record it would seem that Nature made nearly every possible mistake before she reached her greatest achievement Man – or perhaps some would say her worst mistake of all. At one time she put her trust in armaments and gigantic size. Frozen in the rock is the evidence of her failures to provide a form fitted to endure and dominate – failures which we are only too ready to imitate. At last she tried a being of no great size, almost defenceless, defective in at least one of the more important sense – organs; one gift she bestowed to save him from threatened extinction – a certain stirring, a restlessness, in the organ called the brain.

And so we come to Man.

II

It is with some such thoughts as these of the relation of Man to the visible universe that the scientifically minded among us approach the problem of his relation to the Unseen World. It is not with any dogmatic challenge that I have given this outline of evolution. Part of what I have described seems to be securely established; other parts involve a considerable element of conjecture – the best we can do to string together fragmentary knowledge. Scientific theories have blundered in the past; they blunder no doubt to-day; yet we cannot doubt that along with the error there come gleams of a truth for which the human mind is impelled to strive. So brief a summary cannot convey the true spirit and intention of this scientific probing of the past, any more than the spirit of history is conveyed by a table of dates. We seek the truth; but if some voice told us that a few years more would see the end of our journey, that the clouds of uncertainty would be dispersed, and that we should perceive the whole truth about the physical universe, the tidings would be by no means joyful. In science as in religion the truth shines ahead as a beacon showing us the path; we do not ask to attain it; it is better far that we be permitted to seek.

I daresay that most of you are by no means reluctant to accept the scientific epic of the Creation, holding it perhaps as more to the glory of God than the traditional story. Perhaps you would prefer to tone down certain harshnesses of expression, to emphasise the forethought of the Creator in the events which I have called accidents. I would not venture to say that those who are eager to sanctify, as it were, the revelations of science by accepting them as new insight into the divine power are wrong. But this attitude is liable to grate a little on the scientific mind, forcing its free spirit of inquiry into one predetermined mode of expression; and I do not think that the harmonising of the scientific and the religious outlook on experience is assisted that way. Perhaps our feeling on this point can be explained by a comparison. A business man may believe that the hand of Providence is behind his commercial undertakings as it is behind all the vicissitudes of his life; but he would be aghast at the suggestion that Providence should be entered as an asset in his balance sheet. I think it is not irreligion but a tidiness of mind, which rebels against the idea of permeating scientific research with a religious implication.

Probably most astronomers, if they were to speak frankly, would confess to some chafing when they are reminded of the psalm "The heavens declare the glory of God." It is so often rubbed into us with implications far beyond the simple poetic thought awakened by the splendour of the star-clad sky. There is another passage from the Old Testament that comes nearer to my own sympathies:

> "And behold the Lord passed by, and a great and strong wind rent the mountains, and brake in pieces the rocks before the Lord; but the Lord was not in the wind: and after the wind an earthquake; but the Lord was not in the earthquake: and after the earthquake a fire; but the Lord was not in the fire: and after the fire a still small voice.... And behold there came a voice unto him, and said, What doest thou here, Elijah?"

Wind, earthquake, fire – meteorology, seismology, physics – pass in review, as we have been reviewing the natural forces of evolution; the Lord was not in them. Afterwards, a stirring, an awakening in the organ of the brain, a voice which asks "What doest thou here?"

III

We have busied ourselves with the processes by which the electric particles widely diffused in primeval chaos have come together to build the complexity of a human being; we cannot but acknowledge that a human being involves also something incommensurable with the kind of entities we have been treating of. I do not mean to say that consciousness has not undergone evolution; presumably its rudiments exist far down the scale of animal life. But it is a constituent or an aspect of reality which our survey of the material world leaves on one side. Hence arises insistently the problem of the dualism of spirit and matter. On the one side there is consciousness stirring with activity of thought and sensation; on the other side there is a material brain, a maelstrom of scurrying atoms and electric charges. Incommensurable as they are, there is some kind of overlap or contact between them. As the mind is traversed by a certain thought the atoms at some point of the brain range themselves so as to start a material impulse transmitting the mental command to a muscle; or again a nervous impulse arrives from the outer world, and as the atoms of a brain-cell move in response to the physical forces simultaneously a sensation of pain occurs in the mind.

Let us for a moment consider the most crudely materialistic view of this connection. It would be that the dance of atoms in the brain really constitutes the thought, that in our search for reality we should replace the thinking mind by a system of physical objects and forces, and that by so doing we strip away an illusory part of our experience and reveal the essential truth which it so strangely disguises. I do not know whether this view is still held to any extent in scientific circles, but I think it may be said that it is entirely out of keeping with recent changes of thought as to the fundamental principles of physics. Its attractiveness belonged to a time when it was considered that the way to understand or explain a scientific phenomenon was to make a concrete mechanical model of it.

I cannot in a few moments make clear a change of thought which it has taken a generation to accomplish. I can only say that physical science has turned its back on all such models, regarding them now rather as a hindrance to the apprehension of the truth behind the phenomena. We have the same desire as of old to get to the bottom of things, but the ideal of what constitutes a scientific explanation has changed almost beyond recognition. And if to-day you ask a physicist what he has finally made out the æther or the electron to be, the answer will not be a description in terms of billiard balls or fly-wheels or anything concrete; he will point instead to a number of symbols and a set of mathematical equations which they satisfy. What do the symbols stand for? The mysterious reply is given that physics is indifferent to that; it has no means of probing beneath the symbolism. To understand the phenomena of the physical world it is necessary to know the equations which the symbols obey but not the nature of that which is being symbolised. It would be irrelevant here to defend this change, to make clear the intellectual satisfaction afforded by these symbolic equations, or to explain why the demand of the layman for a concrete explanation has to be set aside. We have, however, to see how this newer outlook has modified the challenge from the material to the spiritual world.

For those who were bent on finding a model for everything, the material brain appeared in the light of a ready-made model of the mind. And being a model, it was for them the full explanation of the mind. A mechanism of concrete particles, like the billiard-ball atoms of the brain, was their ideal of an explanation. They were hoping similarly to find a mechanism of gyrostats and cog-wheels to explain the æther. The cog-wheels of the æther were hidden, but the cog-wheels of the mind seemed to be at any rate partly exposed. The mere sight of such machinery gave them a feeling of satisfaction, even if they could not tell in the least how it worked. I am not here greatly concerned with the question whether, or to what extent, the brain-cells may rightly be regarded as the cog-wheels of the mind. What I wish to point out is that we no longer have the disposition which, as soon as it scents a piece of mechanism, exclaims "Here we are getting to bedrock. This is what things should resolve themselves into. This is ultimate reality." Physics to-day is not likely to be attracted by a type of explanation of the mind which it would scornfully reject for its own æther.

Perhaps the most essential change is that we are no longer

tempted to condemn the spiritual aspects of our nature as illusory because of their lack of concreteness. We have travelled far from the standpoint which identifies the real with the concrete. Even the older philosophy found it necessary to admit exceptions; for example, time must be admitted to be real, although no one could attribute to it a concrete nature. Nowadays time might be taken as typical of the kind of stuff of which we imagine the physical world to be built. Physics has no direct concern with that feeling of "becoming" in our consciousness which we regard as inherently belonging to the nature of time, and it treats time merely as a symbol; but equally matter and all else that is in the physical world have been reduced to a shadowy symbolism.

We all share the strange delusion that a lump of matter is something whose general nature is easily comprehensible whereas the nature of the human spirit is unfathomable. But consider how our supposed acquaintance with the lump of matter is attained. Some influence emanating from it plays on the extremity of a nerve, starting a series of physical and chemical changes which are propagated along the nerve to a brain cell; there a mystery happens, and an image or sensation arises in the mind which cannot purport to resemble the stimulus which excites it. Everything known about the material world must in one way or another have been inferred from these stimuli transmitted along the nerves. It is an astonishing feat of deciphering that we should have been able to infer an orderly scheme of natural knowledge from such indirect communication. But clearly there is one kind of knowledge which cannot pass through such channels, namely knowledge of the intrinsic nature of that which lies at the far end of the line of communication. The inferred knowledge is a skeleton frame, the entities which build the frame being of undisclosed nature. For that reason they are described by symbols, as the symbol x in algebra stands for an unknown quantity.

The mind as a central receiving station reads the dots and dashes of the incoming nerve-signals. By frequent repetition of their call-signals the various transmitting stations of the outside world become familiar. We begin to feel quite a homely acquaintance with 2LO and 5XX. But a broadcasting station is not *like* its call signal; there is no commensurability in their nature. So too the chairs and tables around us which broadcast to us incessantly those signals which affect our sight and touch cannot in their nature be like unto the signals or to the sensations which the signals awake at the end of their journey.

Penetrating as deeply as we can by the methods of physical investigation into the nature of a human being we reach only symbolic description. Far from attempting to dogmatise as to the nature of the reality thus symbolised, physics most strongly insists that its methods do not penetrate behind the symbolism. Surely then that mental and spiritual nature of ourselves, known in our minds by an intimate contact transcending the methods of physics, supplies just that interpretation of the symbols which science is admittedly unable to give. It is just because we have a real and not merely a symbolic knowledge of our own nature that our nature seems so mysterious; we reject as inadequate that merely symbolic description which is good enough for dealing with chairs and tables and physical agencies that affect us only by remote communication.

In comparing the certainty of things spiritual and things temporal, let us not forget this – Mind is the first and most direct thing in our experience; all else is remote inference.

That environment of space and time and matter, of light and colour and concrete things, which seems so vividly real to us is probed deeply by every device of physical science and at the bottom we reach symbols. Its substance has melted into shadow. None the less it remains a real world if there is a background to the symbols – an unknown quantity which the mathematical symbol x stands for. We think we are not wholly cut off from this background. It is to this background that our own personality and consciousness belong, and those spiritual aspects of our nature not to be described by any symbolism or at least not by symbolism of the numerical kind to which mathematical physics has hitherto restricted itself. Our story of evolution ended with a stirring in the brain-organ of the latest of Nature's experiments; but that stirring of consciousness transmutes the whole story and gives meaning to its symbolism. Symbolically it is the end, but looking behind the symbolism it is the beginning.

IV

What is the problem that is contemplated when we discuss the possible conflict of the scientific and the religious outlook? I think that so far as the Society of Friends is concerned we should define it as the problem presented by experience – the problem of the proper orientation of our minds towards the different elements of our experience. If science claims in any way to be a guide to life it is because it deals with experience, or part of experience. And if religion is not an attitude towards experience, if it is just a creed postulating an ineffable being who has no contact with ourselves, it is not the kind of religion which our Society stands for. The interaction of ourselves with our environment is what makes up experience. Part of that interaction consists in the sensations associated with impulses coming through our sense- organs; it is by following up this element of experience that we reach the scientific problem of the physical world. But surely experience is broader than this, and the problem of experience is not limited to the interpretation of sense-impressions.

Picture first consciousness as a bundle of sense-impressions and nothing more. As the sensations succeed one another, as they are compared in one consciousness and another, from somewhere comes the query "What are we to think of it all? What is it all about?" To answer this is the purpose of science. But picture again consciousness, not this time as a bundle of sense-impressions, but as we intimately know it, responsible, aspiring, yearning, doubting, originating in itself such impulses as those which urge the scientist on his quest for truth. "What are we to think of it all? What is it all about?" This time the answer must be broader, embracing but not limited to the scientific answer.

Normally it is my task to propagate the truths of science, to urge its outstanding importance, and to tread myself the way by which it seeks an understanding of the phenomena which we experience. It is far from my thought to disparage what we gain by this quest. As truly as the mystic, the scientist is following a light; and it is not a false or an inferior light. Moreover the answers given by science have a singular perfection, prized the more because of the long record of toil and achievement behind them. Why then do I not produce one of these scientific answers now? Simply because before giving an answer, it is usual to listen to the question that is put. It is no use having ready a flawless answer if people will not put to you the question it is intended for. So far as I can judge, the kind of question to which I have exposed myself by coming here tonight is, What is the proper orientation of a rational being towards that experience which he so mysteriously finds himself partaking of? What conception of his surroundings should guide him as he sets about the fulfilment of the life bestowed on him? Which of those strivings and feelings which make up his nature are to be nourished, and which rejected as the seed of illusion? The desire for truth so prominent in the quest of science, a reaching out of the spirit from its isolation to something beyond, a response to beauty in nature and art, an Inner Light of conviction and guidance – are these as much a part of our being as our sensitivity to sense-impressions? I have no ready-made answer for these questions. Study of the scientific world cannot prescribe the orientation of something which is excluded from the scientific world. The scientific answer is relevant so far as concerns the sense-impressions interlocked with the stirring of the spirit, which indeed form an important part of the mental content. For the rest the human spirit must turn to the unseen world to which it itself belongs.

Some would put the question in the form "Is the unseen world revealed by the mystical outlook a reality? "Reality is one of those indeterminate words which might lead to infinite philosophical discussions and irrelevancies. There is less danger of misunderstanding if we put the question in the form. "Are we, in pursuing the mystical outlook, facing the hard facts of experience?" Surely we are. I think that those who would wish to take cognisance of nothing but the measurements of the scientific world made by our sense-organs are shirking one of the most immediate facts of experience, namely that consciousness is not wholly, nor even primarily a device for receiving sense-impressions.

We may the more boldly insist that there is another outlook than the scientific one, because in practice a more transcendental outlook is almost universally admitted. I cannot do better than quote a memorable passage from the Swarthmore Lecture by J. S. Hoyland last year:

> "There is an hour of the Indian night, a little before the first glimmer of dawn, when the stars are unbelievably clear and close above, shining with a radiance beyond our belief in this foggy land. The trees stand silent around one with a friendly presence. As yet there is no sound from awakening birds; but the whole world seems to be intent, alive, listening, eager. At such a moment the veil between the things that are seen and the things that are unseen becomes so thin as to interpose scarcely any barrier at all between the eternal beauty and truth and the soul which would comprehend them."

Here is an experience which the "observer" as technically defined in scientific theory knows nothing of. The measuring appliances which he reads declare that the stars are just as remote as they always have been, nor can he find any excuse in his measures for the mystic thought which has taken possession of the mind and dominated the sense-impressions. Yet who does not prize these moments that reveal to us the poetry of existence? We do not ask whether philosophy can justify such an outlook on nature. Rather our system of philosophy is itself on trial; it must stand or fall according as it is broad enough to find room for this experience as an element of life. The sense of values within us recognises that this is a test to be passed; it is as essential that our philosophy should survive this test as that it should survive the experimental tests supplied by science.

In the passage I have quoted there is no direct reference to religious mysticism. It describes an orientation towards nature accepted by religious and irreligious alike as proper to the human spirit – though not to the ideal "observer" whose judgments form the canon of scientific experience. The scientist who from time to time falls into such a mood does not feel guilty twinges as though he had lapsed in his devotion to truth; he would on the contrary feel deep concern if he found himself losing the power of entering into this kind of feeling. In short our environment may and should mean something towards us which is not to be measured with the tools of the physicist or described by the metrical symbols of the mathematician. We cannot argue that because natural mysticism is universally admitted in some degree therefore religious mysticism must necessarily be admitted; but objections to religious mysticism lose their force if they can equally be turned against natural mysticism. If we claim that the experience which comes to us in our silent meetings is one of the precious elements that make up the fullness of life, I do not see how science can gainsay us. Let it pause before rushing in to apply a supposed scientific test; for such a test would go much too far, stripping away from our lives not only our religion but all our feelings which do not belong to the function of a measuring-machine.

In justifying the place of religious experience in human life, we have not to consider it from the point of view of propagating a creed. We do not send missionaries to the blind to persuade them that it will be to their benefit to believe that a world of light and colour exists for other men gifted with eyes. We should not argue with the blind man who maintained that sight was an illusion to which some abnormal people were subject. Therefore in speaking of religious experience I do not attempt to prove the existence of religious experience, any more than in lecturing on optics I should attempt to prove the existence of sight. What I may attempt is to dispel the feeling that in using the eye of the body or the eye of the soul, and incorporating what is thereby revealed in our conception of reality, we are doing something irrational and disobeying the leading of truth which as scientists we are pledged to serve.

V

I have already said that science is no longer disposed to identify reality with concreteness. Materialism in its literal sense is long since dead. But its place has been taken by other philosophies which represent a virtually equivalent outlook. The tendency today is not to reduce everything to manifestations of matter – since matter now has only a minor place in the physical world – but to reduce it to manifestations of the operation of natural law. By "natural law" is here meant laws of the type prevailing in geometry, mechanics, and physics which are found to have this common characteristic – that they are ultimately reducible to mathematical equations. They may also be defined by a less technical property, viz., they are laws which, unlike human law, are never broken. It is this belief in the universal dominance of scientific law which is nowadays generally meant by materialism.

The harmony and simplicity of scientific law appeals strongly to our aesthetic feeling. It illustrates one kind of perfection, such as we might perhaps think worthy to be associated with the mind of God. One of the important questions that we have to face is whether the unseen world is governed by a like scheme of law. I am aware that many religious writers have felt no objection to, and even welcomed, the intrusion of natural law into the spiritual domain. (Probably, however, they are using the term "natural law" in a more elastic sense than that in which the materialist understands it.) Why (they ask) should we insist for ourselves on exemption from a kind of government which as displayed in inorganic nature might be hailed as a manifestation of divine perfection? But I am sure that those who take this view have never understood and faced the meaning of the ideal scheme of scientific law. What they would welcome is not science but pseudo-science. Analogies can be drawn between spiritual and natural phenomena which may serve to press home a moral lesson. For example, one of Kirchoff's famous laws of radiation states that the absorbing power of substances is proportional to the emitting power, so that the best absorbers are also the best emitters. That might make a good text for a sermon. But if ever scientific law makes a serious inroad into the spiritual domain the consequences will not be limited to supplying texts for sermons.

Natural law is not applicable to the unseen world behind the symbols, because it is unadapted to anything except symbols, and its perfection is a perfection of symbolic linkage. You cannot apply such a scheme to the parts of our personality which are not measurable by symbols any more than you can extract the square root of a sonnet. There is a kind of unity between the material and the spiritual worlds – between the symbols and their background – but it is not the scheme of natural law which will provide the cement.

In saying this I am not forgetting the likelihood of great future developments of science which may and indeed must bring to light types of natural law of which as yet we have no conception. Thus I do not judge the problem of life (in so far as it can be dissociated from consciousness) to be impregnable to the attack of physics. It is a matter of keen controversy among biochemists whether physics and chemistry as they stand are adequate to deal with the properties of living organisms. I express no opinion; but, in any case, whether they are adequate or not today, I cannot assume that future revolutions of science and the admission of new fundamental conceptions will not make them adequate. It is when life is associated with consciousness that we reach different ground altogether. To those who have any intimate acquaintance with the laws of chemistry and physics the suggestion that the spiritual world could be ruled by laws of allied character is as preposterous as the suggestion that a nation could be ruled by laws like the laws of grammar. The essential difference, which we meet in entering the realm of spirit and mind, seems to hang round the word "Ought."

This limitation of natural law to a special domain would be more obvious but for a confusion in our use of the word law. In human affairs it means a rule, fortified perhaps by incentives or penalties, which may be kept or broken. In science it means a rule which is never broken; we suppose that there is something in the constitution of things which makes its non-fulfilment an impossibility. Thus in the physical world what a body does and what a body ought to do are equivalent; but we are well aware of another domain where they are anything but equivalent. We cannot get away from this distinction. Even if religion and morality are dismissed as illusion, the word "Ought" still has sway. The laws of logic do not prescribe the way our minds think; they prescribe the way our minds ought to think.

Suppose we concede the most extravagant claims that might be made for natural law, so that we allow that the processes of the mind are governed by it; the effect of this concession is merely to emphasise the fact that the mind has an outlook which transcends the natural law by which it functions. If, for example, we admit that every thought in the mind is represented in the brain by a characteristic configuration of atoms, then if natural law determines the way in which the configurations of atoms succeed one another it will simultaneously determine the way in which thoughts succeed one another in the mind. Now the thought of "7 times 9" in a boy's mind is not seldom succeeded by the thought of "65". What has gone wrong? In the intervening moments of cogitation everything has proceeded by natural laws which are unbreakable. Nevertheless we insist that something has gone wrong. However closely we may associate thought with the physical machinery of the brain, the connection is dropped as irrelevant as soon as we consider the fundamental property of thought – that it may be correct or incorrect. The machinery cannot be anything but correct. We say that the brain which produces "7 times 9 are 63" is better than the brain which produces "7 times 9 are 65"; but it is not as a servant of natural law that it is better. Our approval of the first brain has no connection with natural law; it is determined by the type of thought which it produces, and that involves recognising a domain of the other type of law – laws which ought to be kept, but may be broken. Dismiss the idea that natural law may swallow up religion; it cannot even tackle the multiplication table single-handed.

VI

Let me play the role of materialist philosopher a few moments longer. The electric particles in obedience to the laws of physics have come together and built human brains. Still in obedience to those laws, they have by their evolutions brought about and stored in those brains the thoughts that make up the sum of human knowledge. Those unbreakable laws have decreed that to-night some of that accumulated knowledge is to be unloosed on you in the form of a lecture. I must hope that you too will be good materialists and feel a due interest in the phenomenon that is proceeding, observing the curious effects of Maxwell's laws, the laws of thermodynamics and other physical causes that are leading to the emission of a modulated system of sound-waves. But no; I was forgetting. That is how as materialists you *ought* to think of my lecture; but "ought" is outside natural law. I cannot expect more than that your brains will react towards the lecture in accordance with the unbreakable laws which govern them; and those who happen to fall asleep may claim that it was decreed by those laws.

This is, of course, a very old *reductio ad absurdum*; and he would be a very shallow materialist who has not appreciated the difficulty and persuaded himself that he has found an answer to it. I am not very curious as to how he surmounts the difficulty or whether his justification is valid. The upshot is that he connives at an attitude towards knowledge which does not treat it as something secreted in the brain by the operation of unbreakable laws of nature. It is to be judged in relation to its truth or untruth not in relation to any supposed theory of its origin.

Truth and untruth belong to the realm of significance and values. I am not able to agree entirely with the assertion commonly made by scientific philosophers that science, being solely concerned with correct and colourless description, has nothing to do with significances and values. If it were literally true, it would mean that, when the significance of our lives and of the universe around us is under discussion, science is altogether dumb. But there is this much truth in it. If we are to present science as a self-contained scheme, owing nothing to any judgements we may have formed by methods for which science does not take responsibility, then no doubt significances must be ruled outside its scope. This may be called the official attitude of science. Officially the scientist is just an adept at solving certain problems; he has no curiosity as to how these problems have come to be set; it is a complete surprise to him that mankind struggling after the eternal verities should take serious note of his pastime. But I think no one would venture to speak to a public audience on any scientific topic unless he were prepared to transgress beyond the official attitude. Imagine a speaker on evolution presenting a purely colourless description of the sequence of living forms and the struggle for existence, without ever hinting at an underlying significance for us of this change in our belief as to Man's place in nature.

The religious seeker who pursues significances and values is often compared unfavourably with the scientist who pursues atoms and electrons. The plain matter-of-fact person is disposed to think that the former is wandering amid shadow and illusion, whilst the latter is coming to grips with reality. I want therefore to give an illustration which will show that unless we pay attention to significances as well as to physical entities we may miss the essential part of experience.

Let us suppose that on November 11th a visitor from another planet comes to the Earth in order to observe scientifically the phenomena occurring here. He is especially interested in the phenomena of sound, and at the moment he is occupied in observing the rise and fall of the roar of traffic in a great city. Suddenly the noise ceases, and for the space of two minutes there is the utmost stillness; then the roar begins again. Our visitor, seeking a scientific explanation of this, may perhaps recall that on another occasion he witnessed an apparently analogous phenomenon in the kindred study of light. It was full daylight, but there came a quick falling of darkness which lasted about two minutes, after which the light came back again. The latter occurrence (a total eclipse of the sun) has a well-known scientific explanation and can indeed be predicted many years in advance. I am assuming that the visitor is a competent scientist; and though he might at first be misled by the resemblance, he would soon find that the cessation of sound was a much more complicated phenomenon than the cessation of light. But there is nothing to suggest that it was outside the operation of the same kind of natural forces. There was no supernatural hushing of sound. The noise ceased because the traffic stopped; each car stopped because a brake applied the necessary friction ; the brake was worked mechanically by a pedal ; the pedal by a foot ; the foot by a muscle ; the muscle by mechanical or electrical impulses travelling along a nerve. The stranger may well believe that each motion has its physical antecedent cause which can be carried back as far as we please; and if the prediction of the two – minute silence on Armistice Day is not predictable like an eclipse of the sun, it is only because of the difficulty of dealing with the configurations of millions of particles instead of with a configuration of three astronomical bodies.

I do not myself think that the intermission of sound was predictable solely by physical laws. It might have been foreseen some days in advance if the visitor had access to the thoughts floating in human minds, but not from any study however detailed of the physical constituents of human brains. I think I am right in saying that within the last two years there has been a change in scientific ideas which makes this more likely than the old deterministic view. But here I am going to grant our visitor his claim; to concede that even human actions are predictable by a – possibly enlarged – scheme of physical law. What then? Shall we let our visitor go away convinced that he has got to the bottom of the phenomenon of Armistice Day? He understands perfectly why there is a two – minute silence; it is a natural and calculable result of the motion of a number of atoms and electrons following Maxwell's equations and the laws of conservation. It differs only from a similar optical event of a two-minute eclipse in being more complicated. Our visitor has apprehended the reality underlying the silence, so far as reality is a matter of atoms and electrons. But he is unaware that the silence has also a significance.

Often the best way to turn aside an attack is to concede it. The more complete the scientific explanation of the silence the more irrelevant that explanation becomes to our experience. When we assert that God is real, we are not restricted to a comparison with the reality of atoms and electrons. If God is as real as the shadow of the Great War on Armistice Day, need we seek further reason for making a place for God in our thoughts and lives? We shall not be concerned if the scientific explorer reports that he is perfectly satisfied that he has got to the bottom of things without having come across either.

VII

We want an assurance that the soul in reaching out to the unseen world is not following an illusion. We want security that faith, and worship, and above all love, directed towards the environment of the spirit are not spent in vain. It is not sufficient to be told that it is good for us to believe this, that it will make better men and women of us. We do not want a religion that deceives us for our own good. There is a crucial question here; but before we can answer it, we must frame it.

The heart of the question is commonly put in the form "Does God really exist?" It is difficult to set aside this question without being suspected of quibbling. But I venture to put it aside because it raises so many unprofitable side issues, and at the end it scarcely reaches deep enough into religious experience. Among leading scientists to-day I think about half assert that the æther exists and the other half deny its existence; but as a matter of fact both parties mean exactly the same thing, and are divided only by words. Ninety-nine people out of a hundred have not seriously considered what they mean by the term "exist" nor how a thing qualifies itself to be labelled real. A late colleague of mine, Dr. MacTaggart, wrote a two-volume treatise on "The Nature of Existence" which may possibly contain light on the problem, though I confess I doubt it. Theological or anti-theological argument to prove or disprove the existence of a deity seems to me to occupy itself largely with skating among the difficulties caused by our making a fetish of this word. It is all so irrelevant to the assurance for which we hunger. In the case of our human friends we take their existence for granted, not caring whether it is proven or not. Our relationship is such that we could read philosophical arguments designed to prove the non-existence of each other, and perhaps even be convinced by them – and then laugh together over so odd a conclusion. I think that it is something of the same kind of security we should seek in our relationship with God. The most flawless proof of the existence of God is no substitute for it; and if we have that relationship the most convincing disproof is turned harmlessly aside. If I may say it with reverence, the soul and God laugh together over so odd a conclusion.

For this reason I do not attach great importance to the academic type of argument between atheism and deism. At the most it may lead to a belief that behind the workings of the physical universe there is need to postulate a universal creative spirit, or it may be content with the admission that such an inference is not excluded. But there is little in this that can affect our human outlook. It scarcely amounts even to a personification of Nature; God is conceived as an all-pervading force, which for rather academic reasons is not to be counted among forces belonging to physics.

Nor does this pantheism awake in us feelings essentially different from those inspired by the physical world – the majesty of the infinitely great, the marvel of the infinitely little. The same feeling of wonder and humility which we feel in the contemplation of the stars and nebulae is offered as before; only a new name is written up over the altar. Religion does not depend on the substitution of the word "God" for the word "Nature."

The crucial point for us is not a conviction of the existence of a supreme God but a conviction of the revelation of a supreme God. I will not speak here of the revelation in a life that was lived nineteen hundred years ago, for that perhaps is more closely connected with the historical feeling which, equally with the scientific feeling, claims a place in most men's outlook. I confine myself to the revelation implied in the indwelling of the divine spirit in the mind of man.

It is probably true that the recent changes of scientific thought remove some of the obstacles to a reconciliation of religion with science; but this must be carefully distinguished from any proposal to base religion on scientific discovery. For my own part I am wholly opposed to any such attempt. Briefly the position is this. We have learnt that the exploration of the external world by the methods of physical science leads not to a concrete reality but to a shadow world of symbols, beneath which those methods are unadapted for penetrating. Feeling that there must be more behind, we return to our starting point in human consciousness – the one centre where more might become known. There we find other stirrings, other revelations (true or false) than those conditioned by the world of symbols. Are not these too of significance? We can only answer according to our conviction, for here reasoning fails us altogether. Reasoning leads us from premises to conclusion; it cannot start without premises. The premises for our reasoning about the visible universe, as well as for our reasoning about the unseen world, are in the self-knowledge of mind. Obviously we cannot trust every whim and fancy of the mind as though it were indisputable revelation; we can and must believe that we have an inner sense of values which guides us as to what is to be heeded, otherwise we cannot start on our survey even of the physical world. Consciousness alone can determine the validity of its convictions. "There shines no light save its own light to show itself unto itself."

The study of the visible universe may be said to start with a determination to use our eyes. At the very beginning there is something which might be described as an act of faith – a belief that what our eyes have to show us is significant. I think it can be maintained that it is by an analogous determination that the mystic recognises another faculty of consciousness, and accepts as significant the vista of a world outside space and time that it reveals. But if they start alike, the two outlets from consciousness are followed up by very different methods; and here we meet with a scientific criticism which seems to have considerable justification. It would be wrong to condemn alleged knowledge of the unseen world because it is unable to follow the lines of deduction laid down by science as appropriate to the seen world; but inevitably the two kinds of knowledge are compared, and I think the challenge to a comparison does not come wholly from the scientists. Reduced to precise terms, shorn of words that sound inspiring but mean nothing definite, is our scheme of knowledge of what lies in the unseen world, and of its mode of contact with us, at all to be compared with our knowledge (imperfect as it is) of the physical world and its interaction with us? Can we be surprised that the student of physical science ranks it rather with the vague unchecked conjectures in his own subject, on which he feels it his duty to frown? It may be that, in admitting that the comparison is unfavourable, I am doing an injustice to the progress made by systematic theologians and philosophers; but at any rate their defence had better be in other hands than mine.

Although I am rather in sympathy with this criticism of theology, I am not ready to press it to an extreme. In this lecture I have for the most part identified science with physical science. This is not solely because it is the only side for which I can properly speak, but because it is generally agreed that physical science comes nearest to that complete system of exact knowledge which all sciences have before them as an ideal. Some fall far short of it. The physicist who inveighs against the lack of coherence and the indefiniteness of theological theories, will probably speak not much less harshly of the theories of biology and psychology. They also fail to come up to his standard of methodology. On the other side of him stands an even superior being – the pure mathematician – who has no high opinion of the methods of deduction used in physics, and does not hide his disapproval of the laxity of what is accepted as proof in physical science. And yet somehow knowledge grows in all these branches. Wherever a way opens we are impelled to seek by the only methods that can be devised for that particular opening, not over-rating the security of our finding, but conscious that in this activity of mind we are obeying the light that is in our nature.

VIII

I have said that the science of the visible universe starts with a determination to use our eyes; but that does not mean that the primary use of the eye is for advancing science. If in a community of the blind one man suddenly received the gift of sight, he would have much to tell which would not be at all scientific. Can we imagine him attempting to convey to his neighbours the significance of the new revelation by talking about the so-called physical "realities"? We know through science that the differences of colour in the external world – red, green, blue – are simply differences of electromagnetic wave-length; and the existence of colour-blindness shows how subjective the effects of the waves on our senses may be. But to the man who has received the revelation of sight the significant fact is not so much the truth about wave-length as the amazing transformation into a world of colour under the vivifying power of the mind. I need not stress the bearing of this when the eye of the soul is opened to an apprehension of the unseen world. The need for expression will not satisfy itself in preaching a scientific sermon. In the world, seen or unseen, there is place for adventure as well as for triangulation; It is right that we should, as far as may be, systematise and criticise the inferences that may be drawn as to the nature of the spiritual world beyond our consciousness; but whatever its abstract frame may be, it is transformed into a different significance when it comes into relation with our consciousness – even as the skeleton frame of scientific truth is transformed into the colour and activity and substance of our familiar environment.

It seems right at this point to say a few words in relation to the question of a Personal God. I suppose every serious thinker is rather afraid of this term which might seem to imply that he pictures the deity on a throne in the sky after the manner of medieval painters. There is a tendency to substitute such terms as "omnipotent force" or even a "fourth dimension." If the idea is merely to find a wording which shall be sufficiently vague, it is somewhat unsuitable for the scientist to whom the words "force" and "dimension" convey something entirely precise and defined. On the other hand, my impression of psychology suggests that the word "person" might be considered vague enough as it stands. But leaving aside verbal questions, I believe that the thought that lies behind this reaction is unsound. It is, I think, of the very essence of the unseen world that the conception of personality should dominate it. Force, energy, dimensions belong to the world of symbols; it is out of such conceptions that we have built up the external world of physics. What other conceptions have we? After exhausting physical methods we returned to the inmost recesses of consciousness, to the voice that proclaims our personality; and from there we entered on a new outlook. We have to build the spiritual world out of symbols taken from our own personality, as we build the scientific world out of the symbols of the mathematician. I think therefore we are not wrong in embodying the significance of the spiritual world to ourselves in the feeling of a personal relationship, for our whole approach to it is bound up with those aspects of consciousness in which personality is centred.

It is difficult to adjust the claims of naïve impressionism and scientific analysis of the spiritual realm without seeming to disparage one or the other; but I think it only requires the same commonsense that we apply to the affairs of ordinary life. Science has an important part to play in our everyday existence, and there is far too much neglect of science; but its intention is to supplement not to supplant the familiar outlook. The biochemist can teach us about the proteins and carbohydrates that make up a suitable diet, and we may profit by his knowledge; but it is not fitting that a meal should be looked upon entirely from the standpoint of absorbing a specified quantity of calories and food-values. It would be still more absurd for a man to refuse food, because he was sceptical as to the certainty of the theories of biochemists. Likewise it is well that there should be some to advise us whether our spiritual bread contains the right kind of vitamins; but for the most part it is the object of our teaching and our meetings to stimulate the spiritual appetite rather than to conduct this kind of research.

If the kind of controversy which so often springs up between modernism and traditionalism in religion were applied to more commonplace affairs of life we might see some strange results. Would it be altogether unfair to imagine something like the following series of letters in our correspondence columns? It arises, let us say, from a passage in an obituary notice which mentions that the deceased had loved to watch the sunsets from his peaceful country home. A writes deploring that in this progressive age few of the younger generation ever notice a sunset; perhaps this is due to the pernicious influence of the teaching of Copernicus who maintains that the sun is really stationary. This rouses B to reply that nowadays every reasonable person accepts Copernicus's doctrine. C is positive that he has many times seen the sun set, and Copernicus must be wrong. D calls for a restatement of belief, so that we may know just how much modern science has left of the sunset, and appreciate the remnant without disloyalty to truth.

E (perhaps significantly my own initial) in a misguided effort for peace points out that on the most modern scientific theory there is no absolute distinction between the heavens revolving round the earth and the earth revolving under the heavens; both parties are (relatively) right. F regards this as a most dangerous sophistry, which insinuates that there is no essential difference between truth and untruth. G thinks that we ought now to admit frankly that the revolution of the heavens is a myth; nevertheless such myths have still a practical teaching for us at the present day. H produces an obscure passage in the Almagest, which he interprets as showing that the philosophy of the ancients was not really opposed to the Copernican view. And so it goes on. And the simple reader feels himself in an age of disquiet, insecurity and dissension, all because it is forgotten that what the deceased man looked out for each evening was an experience and not a creed.

IX

In its early days our Society owed much to a people who called themselves Seekers; they joined us in great numbers and were prominent in the spread of Quakerism. It is a name which must appeal strongly to the scientific temperament. The name has died out, but I think that the spirit of seeking is still the prevailing one in our faith, which for that reason is not embodied in any creed or formula. It is perhaps difficult sufficiently to emphasise Seeking without disparaging its correlative Finding. But I must risk this, for Finding has a clamorous voice that proclaims its own importance; it is definite and assured, something that we can take hold of – that is what we all want, or think we want. Yet how transitory it proves. The finding of one generation will not serve for the next. It tarnishes rapidly except it be preserved with an ever – renewed spirit of seeking. It is the same too in science. How easy in a popular lecture to tell of the findings, the new discoveries which will be amended, contradicted, superseded in the next fifty years! How difficult to convey the scientific spirit of seeking which fulfils itself in this tortuous course of progress towards truth! You will understand the true spirit neither of science nor of religion unless seeking is placed in the forefront.

Religious creeds are a great obstacle to any full sympathy between the outlook of the scientist and the outlook which religion is so often supposed to require. I recognise that the practice of a religious community cannot be regulated solely in the interests of its scientifically-minded members and therefore I would not go so far as to urge that no kind of defence of creeds is possible. But I think it may be said that Quakerism in dispensing with creeds holds out a hand to the scientist. The scientific objection is not merely to particular creeds which assert in outworn phraseology beliefs which are either no longer held or no longer convey inspiration to life. The spirit of seeking which animates us refuses to regard any kind of creed as its goal. It would be a shock to come across a university where it was the practice of the students to recite adherence to Newton's laws of motion, to Maxwell's equations and to the electromagnetic theory of light. We should not deplore it the less if our own pet theory happened to be included, or if the list were brought up to date every few years. We should say that the students cannot possibly realise the intention of scientific training if they are taught to look on these results as things to be recited and subscribed to. Science may fall short of its ideal, and although the peril scarcely takes this extreme form, it is not always easy, particularly in popular science, to maintain our stand against creed and dogma. I would not be sorry to borrow for our scientific pronouncements the passage prefixed to the Advices of the Society of Friends in 1656 and repeated in the current General Advices:

"These things we do not lay upon you as a rule or form to walk by; but that all with a measure of the light, which is pure and holy, may be guided; and so in the light walking and abiding, these things may be fulfilled in the Spirit, not in the letter; for the letter killeth, but the Spirit giveth life."

Rejection of creed is not inconsistent with being possessed by a living belief. We have no creed in science, but we are not lukewarm in our beliefs. The belief is not that all the knowledge of the universe that we hold so enthusiastically will survive in the letter; but a sureness that we are on the road. If our so-called facts are changing shadows, they are shadows cast by the light of constant truth. So too in religion we are repelled by that confident theological doctrine which has settled for all generations just how the spiritual world is worked; but we need not turn aside from the measure of light that comes into our experience showing us a Way through the unseen world.

Religion for the conscientious seeker is not all a matter of doubt and self-questionings. There is a kind of sureness which is very different from cocksureness.

EDDINGTON'S PRINCIPLE IN THE PHILOSOPHY OF SCIENCE

Sir Edmund Whittaker

FRS

THE FIFTH

ARTHUR STANLEY EDDINGTON

MEMORIAL LECTURE

9 August 1951

Eddington's Principle in the Philosophy of Science

MAY I THANK THE Trustees of the Eddington Memorial Fund for the honour they have done me by inviting me to deliver this lecture: and may I also express my pleasure that the invitation has come at a time when Eddington's Fundamental Theory, after making but little progress during the four years since the publication of his posthumous book, has now entered on a new phase of development. Of this advance, which is due to Professor S. R. Milner, F.R.S., I hope to say something later in the lecture.

The last fourteen years of Eddington' s life were spent chiefly in justifying and applying a new Principle which he introduced into the philosophy of science, and which is the central theme of his last book, *Fundamental Theory*. I propose in the present lecture to describe the Principle and so discuss the reasons that may be advanced for and against its acceptance.

The Principle depends on a distinction which Eddington drew between two kinds of assertion that may be made in physics. On the one hand, we may have assertions such as these: 'The masses of the electron, the μ-meson, the π-meson, the τ-meson, and the proton, are approximately in the ratios 1, 200, 300, 1000, 1836': 'The value of the reciprocal of the fine-structure constant is 137': 'The ratio of the electric to the gravitational force between a proton and an electron is 2.2714×10^{39}'. These statements all assert that something has a definite numerical value: they may be called *quantitative* assertions. On the other hand, we may have statements such as: 'A material body which occupies a certain space can also occupy a different space without being changed in its properties': 'The velocity of light is independent of the motion of its source': 'It is impossible to detect a uniform translatory motion, which is possessed by a system as a whole, by means of observations of phenomena taking place wholly within the system': 'Inside a hollow electrified conducting shell there is no electric field': 'It is impossible to derive mechanical effect from any portion of matter by cooling it below the temperature of the coldest of the surrounding objects'. None of these statements mentions any particular number: they may be called *qualitative* assertions. Eddington's Principle depends on the distinction between what we have called quantitative and qualitative assertions: it may be stated thus: *All the quantitative propositions of physics, that is, the exact values of the pure numbers that are constants of science, may be deduced by logical reasoning from qualitative assertions, without making any use of quantitative data derived from observation.*

I am reminded of the adventures of a friend of mine, who, after taking a high place as a Wrangler in the Cambridge Mathematical Tripos, decided that he would then like to take the Natural Science Tripos. As he had never done any physics at school, he had to begin that subject at the beginning; and for his first experiment was asked to determine π, the ratio of the circumference of a circle to its diameter, by cutting out a circle, and also a square whose side was equal to the radius of the circle, in some kind of sheet metal, and then weighing them. He told the demonstrator that mathematicians had much better ways of determining π than this, and indeed they could calculate it to hundreds of places of decimals, whereas any value obtained from sheet metal would be of very limited reliability. This of course was an example of Eddington's Principle: from purely qualitative assumptions, namely the axioms of Euclidean geometry, it was possible to obtain the quantitative value π to any degree of accuracy. This character of absolute precision was exemplified in many of Eddington's own calculations, as for example in the case of the fine-structure constant: no experimental determination of this can be trusted to more than four significant figures, whereas Eddington asserted that it had the exact value 137, with no probable error: and again in the case of the ratio of the electric to the gravitational force between a proton and an electron, his formula yields a result accurate to as many significant figures as are desired.

Let us now consider the historical evolution of the Eddington Principle. The discovery of Archimedes, that π can be calculated by purely theoretical methods, without the necessity for making measurements, shows how far the Greeks had advanced beyond the original conception of geometry (which, as the word indicates, was that of land-measurement or surveying.) No further developments in the direction of Eddington's Principle took place in the Middle Ages or at the Renaissance, when indeed the Baconian doctrines tended to throw all the emphasis on experiment. Leibnitz however initiated a new philosophical outlook: he advocated, in his own words,* a subordination of the science of quantity to the science of quality – of the science that deals with numerical relations to that which treats of order and similarity. This is Eddingtonianism pure and simple.

The first progress in carrying out the Leibnitzian programme was achieved in the field of geometry. Some geometrical theorems which do not involve the notion of distance or measurement had been known in antiquity, such as the theorem of Pappus on the way in which two sets of three collinear points determine a third set of three collinear points; and others were discovered in the seventeenth century by Desargues and Pascal: but these authors did not stress any distinction in principle between their work and metrical geometry: and it was not until 1822 that the study of the projective properties of figures was treated by Poncelet as a more or less self-contained subject. Poncelet did not free himself entirely from dependence on the notion of distance, since he used it in defining the cross-ratios in terms of which projective properties were expressed: but this last link with the old system was broken in 1847, when Karl von Staudt succeeded in defining not only cross-ratios but even distance itself projectively, that is to say, in a com-

* *Leibnizens mathematische Schriften* (ed. Gerhardt, 1863), Vol. VII, *Mathesis Universalis*

pletely qualitative manner. The new projective geometry included the old metrical geometry as a particular case: the qualitative concept of *order* was shown to lead to a greater degree of power and generality than the quantitative concept of *measurement*.

At this stage, remembering that the metrical geometry of Euclid preceded the projective geometry of von Staudt by more than a thousand years, we must consider a first objection to Eddington's Principle, namely that to make the qualitative prior to the quantitative is to put the cart before the horse: for, in the usual account of the development of science, we are told that accurate knowledge regarding the external world begins with the observation and measurement of phenomena that are apparent to our senses, followed by the endeavour to find in the measures some regularity that can be represented by a mathematical formula. Thus the science of optics began when the ancient Greek philosophers examined the path of a ray of light that had been reflected at a mirror, and found by measurement that the angles of incidence and reflection are equal: while the modern history of the subject began when, in the early seventeenth century, Willibrord Snell studied the bending of rays of light as they pass from air into some liquid or solid transparent substance, and discovered that the sines of the angles of incidence and refraction are in a constant ratio.

So far as experimental physics is concerned, it is no doubt true that the performance of a series of measurements of a phenomenon under different circumstances, and the subsequent codification of the measures into a mathematical formula, is the normal procedure. But Eddington's Principle belongs to theoretical, not to experimental physics, and it carries the implication that theoretical and experimental physics are developed by wholly different methods.

Still keeping to the example of the laws of reflection and refraction, we remark that the formulae, though true and important in themselves, are merely mathematical representations of particular phenomena, and do not in themselves provide a starting-point for a general science of optics, capable of uniting into a single theory everything from the colours of thin films and double refraction to the rotation of the plane of polarisation of light by a magnetic field: the real starting-point for the general science was Huygens' introduction of the idea of a travelling wave-front: he assumed that every surface-element of the wave-front at any instant sends out secondary waves, and that the wave-front at any subsequent instant is the envelope of the secondary waves which have arisen from the various surface-elements of the original wave-front. Now it is clear that this Principle of Huygens was not derived from experiment, and the existence of the travelling wave-front cannot be proved experimentally: indeed modern quantum physics has raised difficulties that affect the whole conception. The origin of the Principle is purely intellectual: it belongs to theoretical, not to experimental physics; and it illustrates in a striking way the difference between the two sciences. The point I wish to make now is that Huygens' Principle had a quality that Snell's formula did not possess, namely it could be used to provide explanations, and even to make predictions, of other and quite different phenomena, and to help in the creation of a general science of optics.

Let it then be frankly admitted that a certain body of knowledge must have been created by the methods of experimental physics before theoretical physics can make a start: the formulae of reflection and refraction must be known before Huygens can devise his Principle to explain them; but when the conceptions of theoretical physics have been introduced, they have a vitality of their own, and an adaptability to fields other than those in connexion with which they were introduced; and it is by them that the unification of isolated experimental results into comprehensive general theories is achieved.

The matter is so important from the point of view of the philosophy of science, and in regard to Eddington's Principle, that I will permit myself to illustrate it by another example, namely Clerk Maxwell's introduction of the displacement-current. At the time when Maxwell began his researches, more than ninety years ago, the theoretical physicists were in the habit of explaining phenomena in terms of *aethers*: the luminiferous aether was by then well-established, and it was generally supposed that one or more other aethers were needed to account for the phenomena of electricity, of magnetism, and of gravitation. Maxwell had a metaphysical dislike of filling space several times over with aethers, and set himself to find out if the number could be reduced.

Following the practice of the Cambridge school of physicists, he tried to imagine a mechanical model – an affair with layers of particles rolling on elastic vortex-cells – which would transmit disturbances in the same way as the luminiferous aether, but which would also present features comparable with electric currents, electrostatic force, and magnetism. He obtained differential equations which represented what took place in this model, showing that they might be identified with the known equations of electric and magnetic actions, provided these electromagnetic equations were modified by the insertion of a new term. This insertion amounted to supposing that changes of electric intensity are equivalent to electric currents: a proposition which was unknown to the electricians of the day, and for which no experimental justification could be found. He proved however that if this *displacement-current*, as he called it, were accepted, there would be no difficulty in accounting for all known electric and magnetic actions, and for all optical effects, by means of one and the same aether: so that light could be interpreted as an electromagnetic phenomenon.

The displacement-current, then, was originally a conception of purely intellectual origin, whose only claim to consideration was that it would enable a single aether to perform the functions hitherto assigned to two or more: it was unsupported by any observational evidence, and was strikingly characteristic of theoretical as contrasted with experimental physics. Many years afterwards, two cases were found in which the displacement-current could be regarded as serving the same kind of unifying purpose as had led to its first introduction: these are so typical of the considerations that confirm belief in a hypothesis of theoretical physics that they deserve mention.

The first arose from the discovery in 1879 – the year of Maxwell's death – of a new action of a magnetic field on electric currents – what is generally called the Hall effect. It was almost immediately pointed out that the Hall effect could be connected with the rotation of the plane of polarisation of light by a magnetic field, which had been discovered by Faraday in 1845. In the one case – the Hall effect – the magnetic field acts on an ordinary electric conduction-current in a metal, while in the other case – the Faraday effect – it acts on the Maxwellian displacement-current in a dielectric; but the effect of the magnetic field on the conduction-current in the one case is just the same as its effect on the displacement-current in the other case: the differential equations of the two phenomena are the same, except for the occurrence of the two kinds of current; and the irresistible inference is, that a displacement-current is physically equivalent to an ordinary conduction current, as Maxwell said it was.

The other piece of evidence arose out of J. J. Thomson's discovery in 1893 that any region of free aether, or vacuum, or whatever we like to call it, which is the seat of electric and magnetic fields, is a storehouse of ordinary mechanical momentum. The question is, how does this momentum come into being? In order to generate mechanical momentum, we usually need the action of a ponderomotive force. Now a ponderomotive force of electromagnetic origin does act on a conduction-current, but there is no conduction-current in the free aether. There is however a displacement-current, if the electric field is varying; let us then try the effect of supposing that electromagnetic ponderomotive force acts on a displacement-current in precisely the same way as it acts on a conduction-current and of making also the complementary supposition that an electromagnetic ponderomotive force acts on the magnetic displacement-current which exists in a varying magnetic field. It is found that these assumptions suffice to explain perfectly the existence and amount of the mechanical momentum discovered by Thomson.

In spite of this additional evidence, however, and of the brilliant successes of the electromagnetic theory of light, physicists were reluctant to accept the Maxwellian hypothesis of the displacement-current in the absence of any direct experimental evidence. When I was an undergraduate at Cambridge in the 1890s – that is, thirty years after its discovery – my teachers seemed to be not quite sure about it. So late as 1902, a physicist of real eminence published a book in which many of Maxwell's principles were severely criticised; and Lord Kelvin, who died in 1907, does not seem to have ever been a full believer.

There is to my mind a certain similarity between Maxwell's theory of the displacement-current and Eddington's Fudamental Theory. Each was an intellectual adventure, supported by no direct experimental evidence: though Maxwell was able to point to a successful prediction of the value of the velocity of light, based on his electromagnetic theory, just as Eddington made many predictions of the values of physical constants; but (as I shall point out later) Eddington did not put forward these predictions as the proper tests of the validity of his ideas. In order to judge his system, it is better to turn to work that has been inspired by it, and undoubtedly the work of Professor Milner, to which I referred at the beginning of this lecture, must take first place. I shall therefore say a few words about this.

It is well known that there is one kind of mathematical expressions which is specially appropriate for representing physical quantities, namely the kind known as *tensors*. The reason for their suitability for this purpose depends on a theorem to the effect that if the components of a tensor vanish in any one system of coordinates, they must vanish in every system of coordinates; and therefore the assertion that the components of a certain tensor vanish is a way of stating a physical fact without bringing in such irrelevant matters as the choice of a system of reference.

It has however been known for many years that the existing tensor-calculus is not general enough to comprehend all that should be included in it: there are some physical quantities that cannot be represented by tensors: as Sir Charles Darwin once put it, something has slipped through the net. To overcome this difficulty, the tensor calculus has been supplemented by what is called the spinor calculus; but even with this improvement, the situation was still unsatisfactory.

On the other hand, Eddington based his theory, so far as mathematics is concerned, on what he called *E*-numbers: these are sixteen symbols, on the foundation of which it is possible to construct a non-commutative algebra. But Eddington left the theory of the *E*-numbers in a somewhat unsatisfactory state: they did not fit into the classical theory of tensors, and standard relativity theory was unable to make any use of them. Milner has now formed from them a class of expressions which he calls $\in$-tensors, which may be regarded as a new kind of tensors, since they possess in common with tensors the linearity and the group-property; but they are transformed by rotations of the axes of space and time in a different way from ordinary tensors: and thus he has created a new branch of tensor analysis, which he has used as the basis of new developments in the theories of gravitation and electromagnetism.

At this point it may be observed that there is a notable difference between Theoretical Physics on the one hand, and Pure Mathematics and Experimental Physics on the other, in respect of the enduring validity of the advances that are made. A theorem of Pure Mathematics, once discovered, is true for ever: all the pure mathematics that Archimedes knew more than 2000 years ago is taught without essential change to our students today. And the results of Experimental Physics, so far as they are simply the expression in mathematical language of the unchangeable brute facts of experience, have the same character of permanence. The situation is different with an intellectual adventure such as Theoretical Physics: it is built round conceptions, and the progress of the subject consists very largely in replacing these conceptions by other conceptions which transcend, or even contradict, them.

At the beginning of the century the two theories which seemed most firmly established were that which represented gravitation in terms of action at a distance, and that which represented light as a motion of waves: and we have seen the one in a certain sense supplanted by General Relativity and the other trying to make the best of an uneasy conjunction with quantum-mechanics. The fame of a theoretical physicist rests on the part that his ideas have played in the history of the science: it does not necessarily detract from his importance if none of them survive into the physics of his remote successors.

These considerations suggest a fresh objection to Eddington's Principle: if theoretical physics on its qualitative side has this transitional and impermanent character, how can we expect the Principle to be true? For surely, in view of what has just been said, it would seem to carry the implication that correct quantitative values can be obtained by starting from qualitative conceptions that are bound to be incomplete or incorrect, and are destined to be superseded. Surprising though it may seem, this is what has actually happened in the past. Take for instance the example I mentioned at the beginning of the lecture, the determination of the number π. Archimedes, who first showed how to find it theoretically, believed that Euclidean geometry was the true geometry of the universe, and that the π he discovered was the ratio of the circumference to the diameter of every circle in actual space. Since then we have found reasons to believe that the geometry of the universe is non-Euclidean, and that the ratio of the circumference of a circle to its diameter is *not* equal to π. But the number π still has a meaning; indeed it can be defined without any reference to geometry, as for example in the following way. If two numbers are written down at random, the probability that they will be prime to each other, i.e. will have no common factor, can be shown to be $6/\pi^2$. In a trial when fifty students wrote down five pairs each, it was found that 154 pairs were prime to each other: so from this trial we should get $6/\pi^2 = 154/250$ which gives $\pi = 3.12$. A trial on a much larger scale would doubtless give a much better approximation to the π of Archimedes.

To take another example. The wave conception of light, which certainly fails to account for many optical phenomena, was nevertheless the basis from which innumerable successful predictions were made: among them, one which was decisive at a critical moment in its history. A memoir employing the wave theory was submitted by Fresnel to the French Academy in 1818, when the rival corpuscular theory was dominant. Poisson, who refereed the paper, noticed that Fresnel's theory would indicate the existence of a bright spot at the centre of the shadow of a circular screen: and when this unexpected consequence was found to be verified by experiment, the paper was accepted, and the wave conception came into favour.

Among the qualitative assertions from which, according to Eddington's Principle, our quantitative knowledge of the external world may be deduced, there is however one group which, in contradistinction to the rest of theoretical physics, seems to have a permanent and absolute character; namely the group of what I have elsewhere called *Postulates of Impotence*. These are not hypotheses of a positive nature about the structure of the world around us, but statements of a negative character, to the effect that something is impossible. The oldest of them is the assertion that perpetual motion is impossible, which may be regarded as being in some sense a negative formulation of the law of conservation of energy; and another, which was known in the eighteenth century and was finally established beyond doubt by Faraday, is that it is impossible to set up an electric field in any region of space enclosed by a hollow conductor of any shape and size, by charging the outside of the conductor. It was from this piece of knowledge that Priestley, eighteen years before Coulomb, derived the formula commonly known as Coulomb's law of force between electrostatic charges; and indeed the whole of electric and magnetic science can be derived from the combination of it with another postulate of impotence, namely the postulate of relativity, which asserts the impossibility of detecting a uniform translatory motion, that is possessed by a system as a whole, by means of observations of phenomena taking place wholly within the system. Another well-known postulate of impotence is that no machine can exist which is capable of converting the heat-energy of surrounding bodies, all at the same temperature as itself, into mechanical energy: on this the science of thermodynamics can be based. The cosmological theory of the late Professor E. A. Milne is founded on an assumption which may be stated in the form 'It is impossible to tell where one is in the universe': and modern quantum-mechanics has brought to light some other postulates of impotence, such as that 'When several electrons are present, it is impossible at any instant to assert that a particular one of them is identical with one which has been observed at an earlier instant' and 'It is impossible to measure precisely the momentum of a particle at the same time as a precise measurement of its position is made'.

In the philosophy of science, a postulate of impotence occupies a peculiar position. It is not a direct inference from experiment, such as is met with in experimental physics; nor is it, like the theorems of pure mathematics, a necessary consequence of the structure of the human mind; nor is it again, like most of the hypotheses of theoretical physics, a creation of the free intellect: it is simply the statement of a conviction that all attempts to do a certain thing, however made, are bound to fail. The postulates of this type already known have proved so fertile in yielding positive results – indeed a very large part of modern physics can be deduced from them – that it is not unreasonable to look forward to a time when the entire science can be deduced by syllogistic reasoning from postulates of impotence.

Let us now pass to yet another objection to Eddington's Principle. There is one of its aspects which brings it into conflict with views held by many eminent physicists and astrophysicists at the present time; and as the question is of fundamental importance, I must devote some care to a close examination of it. The point is this: the so-called 'constants of nature', such as the Newtonian constant of gravitation, when calculated in accordance with Eddingtonian ideas, are absolutely constant, that is, they do not vary in time, whereas in several recent cosmological theories they are represented as increasing or decreasing, depending in one way or other on the age of the universe at the time when they are measured.

The first of the constants of nature whose permanent constancy was challenged was the velocity of light: the values obtained experimentally for this quantity by different investigators during the past century seemed to show a small progressive decrease. This fact was however not associated with any theory, and Eddington pointed out that variability of the velocity of light would imply either that the period of light is not determined by the periodicity of its source, or else that the wave-length of light is not determined by the linear scale of its source. There was therefore a general disposition to regard the matter as arising from accidental errors of observation. But other work of Eddington's suggested a further inquiry into the larger question.

Most of the quantities met with in physics, such as the charge or mass of an electron, or the velocity of light, when expressed as numbers, depend on the choice of our units of mass, length, and time; for instance, if we changed the unit of length from a centimetre to an inch, and changed the unit of time from a second to an hour, then the number which stands for the velocity of light would have to be changed. On the other hand, some physical quantities, such as the ratio of the mass of a proton to the mass of an electron are *pure numbers*, that is to say, they are not affected by any alterations of the units of mass, length, and time. The ratio of the circumference of a circle to its diameter, π, is another example of a pure number, since it has the same value whether the circumference and the diameter are measured in inches, feet, or centimetres.

By suitable combination of the different impure numbers that occur in nature it is possible to obtain several that are pure: thus if e denotes the charge of the electron, c denotes the velocity of light, and h represents Planck's constant of action, then $hc/2\pi e^2$ is a pure number, which at the present time has a value that is very nearly, or perhaps exactly, 137. Eddington's theory led him to assert that it is exactly 137 and will remain at this value for all eternity.

Whether it is exactly 137 or not, its permanence in time seems to be assured by observational evidence, as Jordan pointed out; for (recalling that the energy-levels in an atom with any number of electrons can be calculated from a Schrödinger equation, which involves hc/e^2) we can show that if hc/e^2 were not an absolute constant, the lines in the spectrum of an atom in a distant and moving nebula would not differ from the spectrum of an atom that is near us and at rest by the multiplication of all frequencies by one and the same number (the Doppler and Hubble effects), but would be displaced relatively to each other in a complicated way. Since this is not in fact observed, we may conclude that $hc/2\pi e^2$ has a value that is permanent in time.

As to whether this *fine-structure constant*, as it is called, has exactly the value 137, as Eddington said it had, or whether it has some value close to this, it may be recalled that at the time when Eddington's assertion was originally made, the value derived from observation was 137.307, with a probable error of $\pm$ 0.048, so it seemed impossible to reconcile Eddington's prediction with the facts. However, the accepted experimental value has changed in the last twenty years, and it is now 137.009: most physicists are content to speak of it simply as 137.

While the calculation of numerical values for the constants of nature, in accordance with Eddington's Principle, is the feature of his Fundamental Theory that has always attracted most attention, he himself pointed out that this is not the most significant part of it. The aspect of his work to which the greatest weight should be attached is the re-interpretation and reconstruction of qualitative physics, which is the foundation of everything else. I venture to think that too much notice has been taken of the numbers, and too little of the general principles. Eddington's own view is stated clearly in the Introduction to his book *Relativity Theory of Protons and Electrons*: he says explicitly that the theory does not rest on observational tests. The numbers found by the experimenter must be accepted as correct, since they are furnished directly by nature; and the numbers found by his own theory must also be accepted as correct, since they are obtained by purely logical deduction from epistemological principles which he regards as unquestioned. If in any case the empirical and theoretical numbers do not agree, the proper inference is, not that either of them is erroneous, but that the experimenter and Eddington are not talking about quite the same thing. How easily this might happen is easier to understand now than would have been possible a few years ago. It is now known that the values of the charge e and the mass m of an electron which are determined by observation are slightly different from the values of e and m that must be substituted in the fundamental mathematical equations of electromagnetic theory. The development of quantum electrodynamics has in fact shown the necessity for what is called *renormalisation*, which is precisely a recognition of this difference between the observed and the theoretical values of e and m. Now it is to be expected that the values of e and m yielded by Eddington's theory will be the theoretical values that should be substituted in the mathematical equations; and therefore if a small discrepancy is found between Eddington's theoretical value of the fine-structure constant and the value derived from observation, we need not be surprised, or infer that either of them is, properly speaking, in error. Although Eddington did not live to see the development of the modern practice of renormalising e and m, he foresaw that a situation of this kind must arise, and he uttered a warning against expecting too close an agreement between his theoretically-calculated values and those obtained by measurement.

After finding the number 137, Eddington calculated the ratio of the electric force between a proton and an electron to the gravitational force between them, which is at present a number of the order 10^{39}, and also the ratio of the mass of the universe to the mass of a proton, which is at present a number of the order 10^{78}. These two numbers illustrate a curious property of the pure numbers that occur in physics and cosmology, namely that if all the known pure numbers that can be derived from physical quantities are arranged in order of magnitude, they are not scattered in a loose uniformity over the whole range of numbers, but are clustered together in three compact groups which are very widely separated from each other. One of these groups consists of numbers between unity and 1900: the second consists of numbers of the order 10^{39}: and the third consists of numbers of the order 10^{78}. Now 10^{39} is the square root of 10^{78}: and Eddington devined that the whole order of nature is built round a number of the order 10^{78}, which he called the *cosmical number*. Thus he was able to explain the three groups of pure numbers: those of order 10^{78} represent those that contain the cosmical number as a factor: those of order 10^{39} contain the square root of the cosmical number as a factor: and those between 1 and 1900 are those that do not involve the cosmical number at all.

The cosmical number is evidently of paramount importance in the structure of the universe. The question is, what does it represent?

Eddington answered this question by showing that, on his theory the cosmical number must be equal to the total number of independent quadruple wave-functions that satisfy certain conditions, and that its value must be $\frac{3}{2} . 136.2^{256}$ which actually is of the order 10^{78}. This number is of course unchangeable in time. In terms of it he expressed the ratio of the electric to the gravitational force between a proton and an electron (which he was now able to give to a thousand places of decimals if required) and all the other pure numbers that are of the orders 10^{39} and 10^{78}.

The permanence in time of the cosmical number was however challenged in 1938 by Dirac. He remarked that if the generally accepted age of the universe is expressed in terms of a unit of time fixed by the constants of atomic theory, we obtain a pure number of the order 10^{39}, that is, a number involving the square root of the cosmical number as a factor. He inferred that the cosmical number is not an absolute constant, permanent in time, but that it increases as the world grows older, being in fact proportional at any instant to the square of the age of the universe at that instant: and this led him to conclude that many, or perhaps all, of the pure numbers that are at present of the order 10^{39} or 10^{78} are actually variable in time, increasing proportionally to the age of the universe or to its square.

Now one of the numbers of the 10^{78} group is the ratio of the estimated mass of the universe to the mass of a proton – what we may call roughly the number of particles in the cosmos. Dirac hesitated to apply his ideas to this case, and he retained the principle of conservation of mass. Jordan however carried out Dirac's idea consistently, and asserted that the total amount of matter in the universe must be increasing. This was a very radical innovation in physical theory: for since mass is equivalent to energy, it seemed to imply that the total amount of energy in the universe is always increasing, and thus to contradict the principle of the conservation of energy. This particular difficulty Jordan overcame by adopting a suggestion made originally by Haas, namely that the gain of energy represented by the augmentation of mass is exactly balanced by the loss of gravitational potential energy which takes place when extended nebulosity contracts into stars: so that the total amount of energy in the universe is invariable.

The suggestion of Jordan, or, as it is now generally called, the hypothesis of continuous creation, is fundamental in the cosmologies of Bondi, Gold, and Hoyle, which are in other respects quite different from Jordan's and which have attracted much well-merited attention in the last two or three years. Variation in the so-called constants of nature is also a feature of the cosmological system of the late Professor E. A. Milne, though he differs from Jordan in believing that the constant of gravitation is not inversely proportional to the age of the universe, but directly proportional to it, so it is increasing instead of decreasing.

It will be seen that the different cosmologies offer a wide variety of alternatives. So far as I know, however, nobody has yet proposed a cosmology based on the assumption that the total amount of matter in the universe is continually diminishing. This would have the recommendation of supplying a very simple picture of the final destiny of the universe. The world as we know it would end by just fading away until there was nothing left.

To return to Eddington, whether we accept his interpretation of the cosmical constant in terms of quadruple wave-functions or not, I think we are compelled to believe that there is a number, at present at the order 10^{78}, which is fundamental in relation to the structure of the universe. The question is whether, with Eddington, we believe this number to be constant in time, or whether, with Dirac and Jordan, we believe it to be increasing. If we believe it to be increasing, then we must believe that all the physical pure numbers of the orders 10^{39} and 10^{78} are increasing, that is, we must believe that the amount of matter in the universe is increasing, and that the constant of gravitation is decreasing: these two things are connected and cannot be separated. Now with regard to the amount of matter in the universe, there is no direct observational evidence one way or the other, but with regard to the constant of gravitation, we are in a more fortunate position.

The possibility of being able to decide this question is based on the circumstance, that the well-known relation which connects the masses with the luminosities of the stars (which, it may be mentioned in passing, is due to Eddington) involves the constant of gravitation. Dr Teller of the University of Chicago, writing in the *Physical Review*, applied the mass-luminosity relation to our sun, and examined his results in the light of the fact that a reasonably steady temperature must have been required for the existence of life on the earth during some hundreds of millions of years. In this way he was led to the conclusion that the constant of gravitation is unlikely to have varied during that time in inverse proportion to the age of the universe. If this is so, it implies that the cosmical number does not vary in time, and therefore that the hypothesis of continuous creation is inadmissible. While it would perhaps be premature to regard Dr Teller's work as conclusive, I think we are justified in saying that the empirical evidence, such as it is, favours the views of Eddington rather than any of the different systems of Dirac, Jordan, Milne, Bondi and Gold, or Hoyle. Everyone of these theories is however of such high intellectual quality and interest that one feels it is a pity they can't all be true.

The conception of a rule of law, in itself timeless, which is intelligible to our minds and which governs all the happenings of the material world, is the spiritual aspect of physical science. We stand in awe before the thought that the intellectual framework of nature is prior to nature herself – that it existed before the material universe began its history – that the cosmos revealed to us by science is only one fragment in the plan of the Eternal.

From the standpoint that we have now attained we can take a more general survey, and consider whether it throws any light on some recent presentations of the philosophy of religion. The philosophical system of my old friend and teacher Alfred North Whitehead is justly regarded as the most important metaphysical achievement of the present century. Whitehead accepts the principle, that in the endeavour to arrive at a philosophy, it is well to begin by first forming a natural philosophy, that is, a philosophy which is concerned with the physical world and is based on the discoveries of theoretical and experimental science: in a further stage, its scope can be enlarged so as to cover the whole of experience and to investigate the general notions of Being and Reality. The completed system is thus a philosophical generalisation of the concepts of science. The development of Whitehead's own thought had this character, from its first communication in his memoir of 1905 *'On mathematical concepts of the material world'* to its consummation as set forth in his Gifford Lectures of 1927 on *Process and Reality*.

Now *Process and Reality* is theistic, in the sense that a Being who is called God occupies a central position in it, and Whitehead affirms that metaphysical principles are just truths about the nature of God. His God, however, bears little resemblance to the Creator revealed in the Book of Genesis, or to the God indicated by the Five Ways of St Thomas Aquinas – the First Mover, the Ultimate Cause, the Necessary Existent. Whitehead's God is not omnipotent; He is not the whole of Reality, and His nature never reaches a static completion, but is always being completed by the creative passage of events: God and the world are, in Whitehead's phrase, instruments of novelty for each other.

The notion of a God who is subject to limitation is a very old one: in the Stoic philosophy of the ancient world, the Supreme Being was conceived as under the sway of Fate: behind the throne of Zeus stood Moira. The emergence in the twentieth century of a doctrine having some resemblance to this leads us to inquire what led Whitehead to adopt it.

I think that its presence in his system may be traced to his way of regarding God and the world-process as bound up with each other, and, so to speak, co-eternal. The difficulties of such a conception become apparent when cosmologists tell us that the age of the material universe is perhaps no more than three thousand million years, and that it is running down like a clock, so that after a certain time it can no longer be the abode of life. The world-process, then, has a beginning and an ending: but God, whatever other qualities He may or may not have, must certainly be eternal: He has neither beginning nor end. The universe known to us, far from being put on the same level of permanence as God, cannot be regarded as more than one of His dispensations: perhaps not the only one.

When the relation of the material universe to God is conceived in this way, it is natural to inquire whether there is in it some meaning for us, which we may hope to understand. A partial answer to this question may be found, I think, when we fix our attention on the order that has been discerned in the cosmos by natural philosophers. The fact that changes in our material universe can be predicted – that they are subject to mathematical law – is the most significant thing about it, for mathematical law is a concept of the mind, and from the existence of mathematical law we infer that our minds have access to something akin to themselves that is in or behind the universe. And the mathematical laws were not brought into existence at the time when they were discovered: men lived in the midst of all the potentialities of nature for countless generations, unable to use them for want of knowledge. The laws of electricity were as true in the age of palaeolithic man as they are today. The natives whom Darwin found at Tierra del Fuego eating raw shellfish off the rocks might have been supported against the rigours of the climate in electrically-heated dressing-gowns if they had known more about laws that were already in existence. What even we know now is only a very small part of the whole that is to be discovered.

In the laws of nature, known and unknown, we recognise a system of truth, which has been revealed to us by the study of nature, but which is unlike material nature in its purely intellectual and universal character, and which, if the conclusions we have reached are correct, is timeless, in contrast to the transitory universe of matter. Material nature has made manifest to our understanding realities greater than itself, realities which point to a God who is not bound up with the world, who is transcendent and subject to no limitation. The principle, that matter exists not for its own sake, but in order to help us in bridging the gulf that separates us from the divine, may be expressed in theological language by saying that nature has a sacramental quality; a principle that has long been recognised in religion, and can now be admitted to be not alien to the philosophy of science.

THE SOURCES OF EDDINGTON'S PHILOSOPHY

Herbert Dingle

DSc ARCS FRAS

THE EIGHTH

ARTHUR STANLEY EDDINGTON

MEMORIAL LECTURE

2 November 1954

The Sources of Eddington's Philosophy

IT IS ALL BUT ten years since the eponymous subject of this series of lectures laid down his labours and left to astronomy new foundations for research, to physics an enigma to solve, to philosophy a source of bewilderment and incredulity, and to the student of human mentality a fascinating field for reflexion. It is to the last two, and especially the last, of these bequests that I invite your attention this evening. As an astronomer Eddington's place is secure. Whatever the future may reveal of the constitution of the stars will be brought to our knowledge because of his pioneering efforts more than a generation ago. His fundamental physical theory, received during its gradual unfolding with an unhappy mixture of incomprehension and scepticism, is viewed with much the same feelings today, but we may be sure that the last word on this matter has not yet been said. On neither of these subjects is comment appropriate at this time, at any rate by me – in the former case because it is superfluous and in the latter because it is impossible. There remains the development of Eddington's general philosophy, and what I hope to do is to throw what light I can on the circumstances that caused it to take the course it did.

I first made contact with the workings of Eddington's mind through his writings, and then through occasional lectures. My first meeting with him, I believe, was when as President he admitted me to Fellowship of the Royal Astronomical Society. Long afterwards I reminded him of this, and he excused himself on the ground that he didn't know what he was doing. This pleased me very much, because I had often told him that but he had never before been brought to confess it. I mention this little incident partly to show that, if we did not always see eye to eye, it was not through lack of frankness on both sides, but chiefly because I think in all seriousness that one of the most significant of Eddington's characteristics was that he didn't know what he was doing, that he had no idea – or, more correctly, a quite false idea – of what his work really implied. He was a creator of high originality, but he was a poor critic. He let his instinct guide him – and what an instinct it was! – but his attempts to rationalize it were either incomprehensible or clearly wrong. I am not going to discuss the age-old problem of the relative importance of creators and critics. You may, if you wish, call the critic a creator *manqué* who could disappear not only with no loss to the world but with liberation of the creative spirit; or you may regard him as the interpreter who brings to the clear light of understanding what the creator achieves blindly, as the engineer comprehends the principles of which the spider, who alone can build the web, knows nothing. Let all that be as it may, the fact remains that Eddington's gift to the world consists in what he did, not in what he said he was doing. His failure to appreciate other people's achievements is evident enough in the few reviews which he reluctantly wrote when it was difficult to avoid doing so, and I believe that the same failure is no less evident in his estimate of his own work.

Why was this so? In the last resort, of course, it is useless to ask why minds are as they are. Let us accept them and be thankful, and leave to the distant future the ultimate problem of their origin and nature. But we can, I think, take a step towards a solution, and the thesis I wish to lay before you can be summarized as follows. Eddington had an exceptional capacity for understanding the background of his subject, which is very rare in those who lead its frontier activities. Consequently, when the theory of relativity burst upon the world he realized fully and with astonishing speed its profoundly revolutionary character. But he was also extremely precocious, and by the time that he realized the significance of relativity, which he tells us was in 1916, he was already thirty-four years old. His view of the nature of the scientific problem was by that time so firmly established in his mind that it was impossible, even for him, then to shake it off; and that view was incompatible with the implications of relativity. Hence all that he could do was to force the two incompatibles into an unnatural union in which the requirements of relativity, though fully accepted, were distorted to the form of a philosophy not flexible enough to receive them in their natural state. To use a metaphor which he adopted when he thought he was describing physicists in general but was really engaged in introspection, he fitted the organism of relativity into the Procrustean bed of his Victorian philosophy. In contemplating the spectacle one hardly knows which feeling prevails – sorrow at the torture or admiration at the achievement. We can imagine what the world would have gained had he been born fifteen years later, but he could hardly have accomplished a mightier work.

To make clear what I mean it is necessary to digress for a while and to consider the nature of the change which relativity demanded in our understanding of the scientific problem. And by relativity I mean what is often called the 'special' theory of relativity, which Einstein published in 1905 but which did not impress itself on Eddington's mind until in 1916 de Sitter pointed out the astronomical consequences of the 'general' theory at which Einstein had then arrived. From a rather narrow mathematical or physical point of view the general theory is, in fact, a generalization of the special theory, for it removes from the consideration of motion the restriction to what is called 'uniformity', but so far as general principles are concerned it is the 'special' theory that contains the essence of the matter. When that theory is once understood and accepted, a generalization to motion of all kinds becomes inevitable, whether or not Einstein's happens to be the right generalization, and it is therefore to the 'special' theory that we must look for light on the fundamental change.

How fundamental that change was is still not generally realized. To the average mathematical physicist, relativity simply consists in substituting the Lorentz for the Galilean transformation of space and time measurements. The 'relativity correction' is a commonly used expression, as though it were simply a matter of putting in some small term that had previously been overlooked. But that is a very superficial view. It usually works for practical purposes, though no one has been more critical of its blind application than Eddington himself – 'it is just muddle-headedness' he once complained to me, in objecting to the work of one of the leaders of theoretical physics. I am concerned now not with the question whether such blind applications are right or wrong, but only with the fact that they are blind. The simple substitution of one mathematical formula for another cannot satisfy the philosophy of science unless its necessary implications are understood; it is not sufficient merely to say that it works. Those who wish to understand the matter must ask themselves why it is the Lorentz and not the Galilean transformation that must be used, and that inevitably leads to the abandonment, not merely of the pre-relativity practices but of the pre-relativity presuppositions also.

To the Victorian physicist, as it is convenient to call him (though the source of his beliefs belongs to an earlier time), the object of science could be described very simply; it was the discovery of the nature of the external world. That world consisted obviously of pieces of matter moving about in time and space. The physicist watched its behaviour under conditions which he could control in his laboratory, and in particular he made measurements because he had learnt by experience that precise relations often existed between the results of measurements of various kinds which were obscured when only the ordinary uncontrolled behaviour of bodies was observed. The measurements were regarded as determinations of the magnitudes of properties possessed by bodies. Thus, each piece of matter possessed a volume, a mass, a density and so on, and instruments, such as meter scales, balances, thermometers, were simply means by which the magnitudes of these properties could be ascertained. But the bodies, with their properties, existed whether you made any measurements or not. An iron bar, for instance, had a length, a mass, a temperature, a density, and many other properties before any measurement of their magnitudes had been made, and would have had them just the same if no one had ever thought of measuring them. The measurement was simply a way of finding out what was already there whether you tried to find it out or not.

The standpoint of relativity implied a complete reversal of all this. You didn't start with the iron rod complete with properties and then discover by measurement what the sizes of those properties were. You started with the measurements that you had made, and gave them names such as length, volume, mass and the rest. The difference may seem merely verbal; in fact it is vital, and goes to the very root of the philosophy of physical science. For consider just one property – length – for simplicity. If this is an inherent property of the rod, you have simply to measure it once for all and then, provided that nothing happens to the rod, you know exactly what it is at any subsequent time. Scientists were all agreed about this; the physicist who looked on the rod as lying at rest in his laboratory, and the astronomer who looked on it as moving at 18 ½ miles a second round the Sun, agreed that this length was whatever the measurement made it out to be. There was no need for each of them to measure it. The length was a property of the rod; it had nothing to do with their view of how the rod was moving.

But if you start with the measurement, and do not use the word 'length' until you have a measurement to bear it, the matter is different. The physicist has to measure the length of a stationary rod and the astronomer the length of a moving one. The recognized procedure for measuring length was quite satisfactory for the physicist, but not so clearly so for the astronomer. If his rod is moving along the measuring scale, he must be careful to read both ends of it at the same time. He has therefore to determine what positions the ends occupy at some single instant, and to his astonishment he found that by no means which he could devise could he do this. There always remained a small measure of uncertainty. In short, there was no procedure known to physics by which the length of a moving rod could be measured.

This, of course, would not in itself have caused any dismay to the Victorian physicist. He could measure the length when the rod was regarded as stationary, and so he could know it; there was no need to repeat under difficult conditions what could be done under easy ones. But unfortunately his researches led him into a contradiction unless the length of a moving rod could be supposed to be less than that of the same stationary one. As we have seen, he could not make the measurement to check this, so he had to *assume* that it was so, that a rod which had a length l when stationary had a length smaller than l when it was moving in the direction of its length. And this he did. He could then still maintain his view that the length of the rod was a property of the world independent of the observer, but he had to suppose that it was not an invariable property, but a property whose magnitude depended on the motion of the rod.

Now this would have been all right if the motion of the rod also were a property of the world independent of the observer; that is to say, if the rod were definitely moving or stationary, quite independently of how you chose to regard it. Clearly there is no inconsistency in saying that one property of the external world depends on another. But this was impossible, for ever since the time of Galileo it had been recognized that if two bodies were in uniform relative motion, there was no objective difference at all between them which would enable you to say which was moving and which, if either, was at rest. This was fundamental to physics, to Victorian as well as twentieth-century physics, because it is enshrined in Newton's first law of motion on which the whole physical scheme is built. And, in fact, the difference between the physicist and the astronomer was not that one was contemplating a stationary and the other a relatively moving rod. They were both contemplating the same rod, in the same state, and the difference between them was simply that one wished to call that state a state of rest and the other wished to call it a state of motion. The assumption that the rod contracted when it moved therefore meant that the rod contracted when you chose to say that it moved, and if two people were in different minds the rod simultaneously had two different lengths.

This was the absurdity which relativity removed. I need not, of course, go into details here, but it will be clear, I think, that since, from the relativity point of view, you were justified in speaking of the length of a stationary rod because you could measure it and so had something to which to give a name, so you could speak of the length of a moving rod if you defined a process of measuring it which could actually be carried out. And this definition the theory of relativity provided. It defined the length of a body *in terms of its motion*, so that it had a definite, ascertainable meaning whatever motion you supposed the body to have, and it was such that, if you supposed the body to have no motion at all, the length was just that which Victorian physics had ascribed to it as an absolute property. In that way the contradiction that had beset Victorian physics was removed.

But that meant, you see, that the whole Victorian view of physics had to be abandoned. I have spoken of length, but in fact not only length but every physical property that a body was supposed to possess had to be redefined so that its value depended on the motion which you were pleased to assign to the body.* The view that physics is the description of the character of an independent external world was simply no longer tenable. Physics became a description of the relations existing between the results of certain operations which you performed, and you chose for yourself what those operations should be. Physical quantities – that is to say, those things that were represented by symbols in physical equations – were not the magnitudes of objective features of the external world. They were the results of your own definitions, and only certain of the relations between them were free from your power to change them by changing your mind. This is not generally

* As special cases, a few properties (e.g. electric charge) happened to be independent of motion – invariant, as we say – but that must be regarded as irrelevant to the general point here discussed. The product of the mass and the volume of a body is invariant, but no one supposes that this has any special significance.

admitted even now,* yet it is inescapable by anyone who accepts the theory of relativity as genuine physics. The argument is too simple to be deniable. Every relativist will admit that if two rods, *A* and *B*, of equal length when relatively at rest, are in relative motion along their common direction, then *A* is longer or shorter than *B*, or equal to it, exactly as you please. It is therefore impossible to evade the conclusion that its length is not a property of either rod; and what is true of length is true of every other so-called physical property. Physics is therefore not the investigation of the nature of the external world.

Now Eddington saw all this with perfect clearness, and marvellously quickly. What it has taken others many years to grasp, he saw almost at once. In what I believe to be his earliest writing on the subject – his Report to the Physical Society on the new general relativity theory, which was published before the eclipse expeditions of 1919 made relativity red-hot news – he states quite clearly that length is not a property of the rod but something which we determine by a physical operation. In his semi-popular book, *Space, Time and Gravitation*, published in 1920, he states still more definitely that where the relativist differs fundamentally from the Victorian physicist is in the hypothesis that 'there is an absolute thing in nature corresponding to length' (p. 7),† and he goes on to say that 'any physical quantity, such as length, mass, force, etc., which is not a pure number, can only be defined as the result arrived at by conducting a physical experiment according to specified rules'. But the most revealing exposition of his view of the whole matter is contained in the Introduction to *The Mathematical Theory of Relativity*, first published in 1923.

I think that much of the bewilderment with which Eddington's later views were received would have been avoided if this Introduction had been studied with sufficient care, for it contains the

---

* E.g. see P. Epstein, *Amer. J. Phys.* 10, 1, 203, 205 (1942), 11, 228 (1943), and M. Born, *Phil. Quart.* 3, 139 (1953).

† All references to Eddington's books are to the first editions.

germ from which all the rest developed. He asserts there (p. 3), and puts the statement in italics, that *'a physical quantity is defined by the series of operations and calculations of which it is the result'.* This is identical with Bridgman's outlook, expressed much later, which has become known as 'operationalism'. It contains also the essence of the contemporary 'logical positivist' thesis, although I think it is in the highest degree unlikely that Eddington had even heard of this school at that time. As is well known, the logical positivists classified all possible statements into three types: synthetic statements, that described the results of physical observations; analytic statements, which were in essence mathematical and tautological; and nonsensical statements. This same analysis, about which volumes have been written, is given in a sentence in Eddington's Introduction (p. 3), where he says that 'any attempt to describe a condition of the world otherwise [than by 'the study of physical quantities'] is either mathematical symbolism or meaningless jargon'. He did not, like the logical positivists, try to purchase precision at the cost of significance by attending only to the *language* of the description (even so, the goods have still not been delivered), but, so far as I know, no one has noticed that the root of his philosophy is the same as that of the logical positivists.

All this is clear enough, but now comes the trouble. The logical consequence of this beginning would have been the inference that the Victorian external world could henceforth be ignored. The evidence for its existence had lain in physical observation; that would have been so obvious to a Victorian that he would hardly have thought of saying it unless questioned on the matter. If, then, the results of physical observation are not, after all, descriptions of the world but simply the consequences of performing certain operations, clearly the evidence for that world, as an object presented for study, vanishes and the only function left for the word 'world', if it is retained, is to denote the system of postulated entities that correlates the observations. In other words, the scientific problem, as seen by the Victorians, is reversed. Instead of starting with a given, unknown world and finding out its nature and character by observation, we start with observations and construct (or infer, if you prefer the word) a world to satisfy them.

But this was just the step that Eddington could not take. His Victorian conviction of the primacy of the external world was so deeply rooted that it survived the destruction of the evidence which alone could justify it. Such cases are by no means unknown in the history of science; Copernicus affords an outstanding example. If the stars move round the Earth, the fact that they all move together makes it impossible to doubt that they are fixed on a sphere, and centuries of belief in the sphere so deduced established its existence as a primary fact that could be taken as given. But if the motion of the stars is only the reflexion of the Earth's rotation, then they must *inevitably* appear to move together, and the sphere becomes a baseless hypothesis. Copernicus, although giving the motion to the Earth, nevertheless went on believing in the starry sphere because its existence was fixed so deeply in his mind that he was unaware that evidence was needed for it.

It was just the same with Eddington. Physical observation failed to reveal the world, but the world was there. What was it like? The answer could only be: we do not know; it is essentially mysterious. But he was loath to believe that there was no relation whatever between physical quantities and the world, so he made the only possible assumption; physical quantities *symbolize* the world in some way. Each of them represents a 'condition of the world'. We could, however, form no conception of the nature of any condition of the world from the corresponding physical quantity or quantities. He gives as an illustration the two ways of expressing the remoteness of a star, by its distance and by its parallax – that is, roughly speaking, the apparent size of the Earth's orbit as seen from the star. These, he says, are alternative representations of the same condition of the world. But if you double the distance, the Earth's orbit will appear only half as great, so the parallax is halved. What, then, happens to the symbolized condition of the world; is it doubled or halved or changed in some quite different way? We do not know. The question is meaningless because the nature of the world is incomprehensible.

The situation, then, is this. Outside us is a world, which for consistency I will always call *the external world*, which is exposed to our observation. We can observe it in various ways, but they all fall into two broad classes, which we may distinguish as *metrical* and *non-metrical*. Physics is concerned only with metrical observations, i.e. with measurements, and it is, of course, only measurements that relativity has shown to be functions of our arbitrarily chosen assumption of movement. This seemed to Eddington to place them in a class apart, and it was only to metrical observations that he assigned a purely symbolical character; certain kinds of non-metrical observations he was willing to believe afforded a direct insight into the nature of the external world. He first expressed this distinction in the following terms: 'I venture to say that the division of the external world into a material world and a spiritual world is superficial, and that the deep line of cleavage is between the metrical and the non-metrical aspects of the world.'* This is a legitimate distinction, whether or not it is fundamental, but unfortunately, in a later work, he identified the realm of the metrical with science instead of with physics, and he wrote: 'The cleavage between the scientific and the extra-scientific domain of experience is, I believe, not a cleavage between the concrete and the transcendental but between the metrical and the non-metrical.'† This had the effect of placing most of biology and psychology outside science, and it has caused a great deal of unfortunate controversy. The question is merely a verbal one, it is true, but if one uses words in a way not normally accepted one must expect to be misunderstood. Some years later, when his book, *The Philosophy of Physical Science* (1939), was passing through the Press, he told me that the choice of that title instead of *The Philosophy of Science* was an attempt to keep me quiet. The attempt was unsuccessful,

---

* *Science, Religion and Reality*, edited by J. Needham (1926), p. 200.

† *The Nature of the Physical World* (1928), p. 275.

but I am glad to have rendered some slight service to the cause of clarity.

I shall return to the question of non-metrical observation of the world, but for the moment let us continue with Eddington's view of physics. Outside us, I repeat, is the inscrutable external world, and we make measurements which symbolize in some unknown way its 'conditions'. We can derive from our measurements the idea of a coherent system which at one time we would have called a description of the world but which now we must call a symbol, a metaphor, in which each element stands for, but in no way resembles, some element of the external world. On account of this detailed correspondence the physical description can be said to indicate the *structure* of the world, just as a map shows the structure of a country, though the graded intensity of colour has no resemblance at all to the variation of height as one crosses a mountain range, and a thin blue line is quite unlike the muddy liquid we may be unlucky enough to fall into. The physical description, which corresponds to the map, we may call the *physical world*, in contrast to the *external world* which is the unknown thing symbolized by it. Eddington does not always use these terms consistently,* and this is one source of his obscurity in spite of his great gift of clear expression. For instance, the word 'electron' is used indiscriminately for the physically defined entity and for its unknown counterpart in the external world. He speaks with confidence of the behaviour of electrons in his description of physical ideas – for example, in his illustration of the uncertainty principle† – and in the next breath tells us that the electron is *'something unknown ... doing we don't know what'*.‡ In one case the 'electron' stands for the dot on the

* In a late book he wrote: 'I usually call X the "external world", the "physical world" being limited to the structure of the external world' (*The Philosophy of Physical Science* (1939), p. 150).

† *The Nature of the Physical World*, p. 224.

‡ Ibid. p. 291.

map, the constituent of the physical world, and in the other it stands for the populous city, the constituent of the external world.

At the risk of tediousness, let me try to make clear the distinction between a *physical quantity*, which, according to Eddington and I think quite properly, is defined by the operations of which it is the result, and what we may call a *physical object*, such as an electron. A physical quantity is anything that is represented by a symbol in a physical equation, and the symbol always stands for the result of a measurement or a group of results of measurements. Thus, the symbol l, standing for the length of a rod, represents a physical quantity, namely, the result of applying to the rod in question the standard procedure for measuring lengths. Some symbols are less directly related to the measuring procedure, but they are always ultimately expressible in terms of it. For instance, the symbol m, standing for the mass of an electron, does not represent the result of applying to an electron the standard procedure for measuring mass. Such a procedure would be intrinsically impossible. Yet we can, in fact, determine m and give it a numerical value, and we do so by making a lot of measurements of ordinary laboratory objects, similar in character to measurements of the length of a rod or the mass of a stone, and combining their results together in a certain way. Strictly speaking, then, we have no right to call m a mass; it is a shorthand symbol for a complicated combination of various measurements of pieces of apparatus in the laboratory. Every symbol we use, and therefore every physical quantity, can be reduced to measurement in this way, and the bare, unadorned meaning of every physical equation is simply and solely that various measurements which we make are found to be related to one another in the way expressed by the equation.

Why, then, do we call m the 'mass of an electron'? Simply because we find it convenient (indeed, our minds being what they are, practically indispensable) to form some imaginary picture of what is taking place behind the scenes, so to speak, of our visible experiment. Thus, we might pass a current of electricity (as we express it) through a vacuum tube and observe a patch of green light at a certain place on the tube. On bringing a magnet near, the patch moves to another place. We measure the change of position, the distance of the magnet, and various other things, and combine the measurements in a certain way so as to give a certain result. But then we superpose on all that an imaginary picture of invisible things called 'electrons' which travel along the tube and give it a green glow where they strike it. We imagine also what we call 'lines of magnetic force' streaming out from the magnet and deflecting the electrons so that they strike the tube at a different place. Having added this picture to the actual happenings, we find that the combination of measurements which we have made must correspond to the mass of one of the electrons if we are to maintain it consistently. We therefore write the symbol *m* for that combination and say that we have measured the mass of the electron.* When the same picture is applied to other experiments, we find that it fits, assuming this same mass for the electron, and we can go on adding detail to the picture to make it applicable to one experiment after another. We thus become in time so familiar with the picture that we think of our operations only in terms of it. The directly observable things and the particular measurements that we actually make recede to the back of our minds, and we suppose ourselves to be investigating directly a world of electrons, protons, photons, magnetic and electric fields, and so on, and regard that as *the physical world* which we are studying. We forget that we have ourselves invented that world in order to give a meaning to our measurements of quite other things, and speak as though we had been presented with it in the first place and had then measured its parts.

Now Eddington, as I said, understood all this perfectly well. He knew that the physical quantities were simply the results of measurements and were not properties of particles. He knew that an electron was a part of a conceptual physical world, something

* The experiment has been somewhat simplified, but not distorted in principle.

whose definition was wholly contained in the definitions of the measurements of other, observable, things that had made it possible consistently to conceive it. But he could not rid his mind of another world behind the physical one and symbolized by it – the external world as I have called it – in which the electron 'in itself' existed. It would have been better if he had kept the name 'electron' for the inhabitant of only one world – either the physical or the external one – and chosen a different name for its counterpart in the other world. However, he did not, so we must put up with a description in one place of what is said elsewhere to be indescribable. It represents no confusion in Eddington's own thought, and we shall, I think, understand him correctly if we picture the three distinct entities out of which his philosophy was built: first, physical quantities, which are simply the results of actual measurements, i.e. such things as ordinary measurements of lengths, volumes, masses, etc.; secondly, the physical world, which is an imaginary structure of which the physical quantities can be regarded as affording indirect measures and which comprises the ordinary entities of present-day physics – fundamental particles, fields of force, etc.; and finally, the external world, consisting of entities unknowable in themselves but of which the physical quantities are symbols and of the structure of which the physical world is a representation.

It needs but a brief reflexion on this state of affairs to make it clear that the external world plays no part at all in the business, and could be left out without the loss of anything. De Sitter once remarked that certain masses of matter in Einstein's early model of the universe fulfilled no other purpose than to enable us to suppose them not to exist, but Eddington's external world denies us even that privilege. He will not let us suppose it not to exist. As physicists we must acknowledge it, although physics can tell us nothing about its nature and would be exactly what it is if it were not there at all. It is thus a useless encumbrance. But unfortunately it is far worse than that. In Eddington's philosophy it becomes a will o' the wisp, leading us astray and finally landing us in a bog of nescience from which no escape seems possible. Let us see this process at work in two directions; the first which I will take is his treatment of religion.

Now Eddington was a deeply religious man, and he was specially anxious to harmonize his religion with his scientific convictions. And this, indeed, would have been simple enough if he had adopted without complication the outlook which relativity had made possible in physics. I am far from saying that no points remain at issue between religion and science, but the particular problem that worried Eddington had, in fact, ceased to exist. It originated in the Victorian prejudices, and can be summarized in the following way. The external world, which is the real world and which we explore by physical investigation, has been examined from top to bottom, and it contains nothing that can be regarded as an object of religious experience, nothing that we can call God, nothing whose behaviour is not mechanically determinable, nothing worthy of worship by free, intelligent beings. Hence religion is an illusion, of the same nature as dreams and hallucinations that are universally admitted to be negligible when we undertake to describe the true scheme of things.

This argument depends for its validity on the assumption that the world outside is the primary reality, and our experience valid only if it can be traced to a source in that world. But, from the relativity point of view, experience is the primary thing and the world is constructed or inferred from experience; no statement about it is tenable that cannot be justified by experience. The reason why the Victorian world contained nothing corresponding to religious experience is then obviously because religious experience had not been taken into account in building it up. The religious man, however, has the same title to construct a world in which the source of his religious experience finds a place as the physicist has to construct a world in which the source of his sensory experience finds a place. In the present state of knowledge it may be – indeed, must be – quite a different world from the physical world, but that is merely a sign of the immaturity of our studies. At one time the sources of our thermal experiences – heat, temperature, entropy – were quite distinct from the sources of mechanical experiences-mass, force, space, time – but we now picture a world, a 'thermodynamic' world, in which a single set of concepts does duty for both. We are therefore clearly not entitled to grant the world of the religious mystic any smaller degree of 'reality' or 'significance' or 'validity', or whatever word for ultimate importance you prefer, than the world of the physicist. Both may, and probably will, be modified as physics and theology advance, until they finally unite, but in the meantime they are on the same basic footing.

All this Eddington accepted. 'We have to build the spiritual world out of symbols taken from our own personality,' he wrote, 'as we build the scientific world out of the symbols of the mathematician.'* And that is really all there is to be said about it. There is scope for discussion, of course, as to what description of the 'spiritual world' most properly fits the facts, but that can be left to the theologian; it is quite independent of anything that the physicist may find or do. 'We do not ask', says Eddington,† 'whether philosophy can justify such an outlook [the mystical outlook] on nature. Rather our system of philosophy is itself on trial; it must stand or fall according as it is broad enough to find room for this experience as an element of life.' But, having asserted that we do not ask such a question, he immediately devotes some twenty-five pages to the answering of it. And the reason why he does so is that the old Victorian external world, though it had clearly given up the ghost, continued to dangle its corpse before his eyes, and he could not be satisfied until he had found there the fossilized remains of a real Victorian Great First Cause. 'We want an assurance that the soul in reaching out to the unseen world is not following an illusion', he writes;‡ i.e. we want a philosophical justification for that which we have already realized needs none. 'Feeling that there must be more

* *Science and the Unseen World* (1929), p. 50

† *Science and the Unseen World* (1929), p. 29.

‡ Ibid. p. 42.

behind...', he goes on,* and proceeds to establish the proposition that the external world does indeed contain something 'real' which is apprehensible in spiritual experience – or at least that no one can prove that it doesn't.

This shows, I think, clearly enough the character of the bedevilment of Eddington's philosophy by his inherent predispositions. It was not that they falsified his vision of the new revelation – that was accurate enough – but that they diverted his attention from it towards dead problems, complicating his description of what he saw and confusing the minds of his readers who could not be expected to understand a conflict of which he himself was not conscious. They forced him to look for 'more behind', when the essence of the matter was not behind but in front, and led him at last to banish the roots of religion to the world of the physically unknowable instead of recognizing them where he really knew they were – in that which is known more immediately than any external or physical or spiritual world, in experience itself:

> O world unknowable, we know thee,
> Inapprehensible, we clutch thee.

The same complication can be seen in his purely physical philosophy; for my second example I will consider his tussle with time. There are, of course, some very difficult problems connected with time, and the chief contribution that relativity has made to their solution is to show us clearly that previously we had used the same word for at least two different things: first, the time of our own experience, which appears to be inseparable from consciousness itself, and second, the time that we measure in physics, which is a definable concept just like that of space or mass. Relativity distinguished these two things, showing that the former had simply to be accepted as something to which our constructed or inferred physical world had necessarily to conform, while the latter was essen-

---

* Ibid. p. 45.

tially a part of that world and could be modified and redefined *ad lib* to meet the demands of experience in ways that would formerly have been thought absurd. Eddington again comprehended this with a completeness and a speed that are really astonishing, but what did he make of it? Instead of simply recognizing the distinction he had immediately to place some single entity called 'time' in the unknowable external world, and then to plague himself with the problem of how it made a 'dual entry into our consciousness'.* Its 'real nature', of course, was quite inapprehensible, but since, unlike space, it is represented by two types of symbol, there must be some relation between these two types arising from the fact that they symbolized the same thing. After much searching he found a relation. Entropy, which is a physical quantity, measured in terms of 'physical time' among other things, had the characteristic of the 'time of consciousness' that it always went in one direction. Here, said Eddington, is the connexion. Entropy in some way partakes of both symbolical appearance and the real unknowable time. In physics pure and simple it is one of the ordinary quantities that represent 'conditions of the world', of the same nature as heat and temperature. But it also belongs to the non-metrical concepts, such as beauty and melody,† and so it acts as a sort of link between physics and aesthetics.

I do not think that anyone but Eddington would have had the amazing combination of courage, perverted outlook and persuasive skill necessary to advance such an utterly fantastic notion as this. There have been some remarkable theories of aesthetics, but I do not know of any that has ventured to call on entropy to help it out of its difficulties. Moreover, it was not long after this pronouncement was made that the discovery of the expansion of the universe, which he fully accepted and which his own philosophy made a necessary and not an accidental phenomenon, provided an equally good example of a physical process showing a one-way

* *The Nature of the Physical World*, p. 91.

† Ibid. p. 105.

direction in time, and by the same token this also should have found a place in aesthetics. I am not aware, however, that he ventured to make this claim, and indeed even his effrontery would hardly have been equal to such an outrage. His only comment, so far as I know, was that the expansion of the universe provided only a large-scale criterion of the direction of time and that 'the position of entropy as the unique *local signpost* remains unaffected'.* This, it may be noted, is not true. In the last resort, entropy provides a signpost only for a closed system, and since it is impossible to screen off gravitational action, there is no strictly 'closed system' except the whole universe. Moreover, even if the distinction were a real one it is difficult to see what the scale of the phenomenon has to do with the matter. But Eddington's prejudice was so strong that he could not shake off this weird interpretation of entropy, and he even looked forward 'in the next few years' to the discovery of some hitherto unknown relation between the expansion of the universe and the second law of thermodynamics. I am not aware that this has yet shown itself. But there is another side to the matter, and when we fully realize his problem we hardly know which feeling is the stronger, contempt for his conclusion or admiration for his achievement. We need to place ourselves in his position to appreciate the magnitude of his success. For what he had to do was to reconcile the existence of the Victorian external world with the destruction by relativity of all evidence for it, and he did it. He could neither deny the one nor reject the other. There is an old riddle: what happens when an irresistible force meets an immovable obstacle? The answer is now available: it is Eddington's philosophy.

It would be a fascinating task to follow the development of his general ideas from their first considered expression† to their latest form, but space prevents this, and I will limit myself to two aspects: his treatment of non-metrical experiences and his conclu-

* *New Pathways in Science* (1935), pp. 67–8.

† *Science, Religion and Reality*, pp. 187–218.

sion that the whole of physical law is potentially deducible by reason without recourse to experience. It will be remembered that, in Eddington's view, the fundamental cleavage in the external world was not between the material and the spiritual, but between the metrical and the non-metrical aspects of it. We must first of all understand this distinction, which is not merely the giving of different names to already recognized dissimilarities, but a radically different classification. It is customary to regard such things as chemistry, cookery, sport as pertaining to the material side of things, and art, music, love, religion as wholly spiritual. Eddington rejected this contrast, and divided each of them into a metrical and a non-metrical element. He saw more affinity between the metrical aspects of, say, chemistry and music, than between the metrical and non-metrical aspects of either. The reason, I think, is clear. Before the coming of relativity his Victorian external world was something which sense observation revealed to him, and, like others, he believed that scientific discovery was making its character known. But relativity changed all that. All measures were now dependent on the motion of the observer, and the motion of the observer was not an objective thing, but something that could be assigned quite arbitrarily. Consequently, all measures became arbitrary, and the only things about them that could be called objective were certain mathematical relations. These told you only the *structure* of the external world, nothing at all about its *nature*. Consequently, all metrical investigation led only to a physical world, not to the real external world – that is, to a world which, having the same structure as the external world, revealed it only symbolically. The real nature of the external world was inaccessible by metrical investigation.

But this did not necessarily apply to non-metrical investigation. Relativity said nothing at all about the activities-or at any rate the essential activities – of the artist, the theologian, the biologist. There was therefore no compulsion to give up the belief that these inquirers were actually getting some knowledge of the real nature of the external world. There was equally, of course, no evidence in favour of such a belief. It is, to put it mildly, not obvious that what the artist, the theologian and the biologist reveal is a part of something of which another part has the structure shown by physical theory. However, they do reveal something, and since nothing that exists can possibly be unrelated to the great reality of the external world, what else can be said about them than that in some way they give us an insight into its nature? 'The suggestion is', he wrote,* 'that when we succeed in making progress with the study of the objective† world, the result will be very different from present-day physics, and that there is no particular reason to expect that it will be called physics. We have spoken of this as a development in the future; but may it not have occurred already? It seems to me that the "enlarged" physics which is to include the objective as well as the subjective is just *science*; and the objective, which has no reason to conform to the pattern of systematization that distinguishes present-day physics, is to be found in the nonphysical part of science. We should look for it in the part of biology (if any) which is not covered by biophysics; in the part of psychology which is not covered by psychophysics; and perhaps in the part of theology which is not covered by theophysics. The purely objective sources of the objective element in our observational knowledge have already been named; they are *life, consciousness, spirit*. We reach then the position of idealist, as opposed to materialist, philosophy. The purely objective world is the spiritual world; and the material world is subjective in the sense of selective subjectivism.'

It is not the least of Eddington's peculiarities that he succeeded in detaching biology from physics and grouping it with theology; but again, in his unique situation, what was he to do? His fundamental cleavage between the metrical and the non-metrical was forced on

---

* *The Philosophy of Physical Science*, pp. 68–9.

† Subjective, in Eddington's sense of the word, is that which 'depends on the sensory and intellectual equipment which is our means of acquiring observational knowledge' (*Philosophy of Physical Science*, p. 17). Objective is therefore that which is independent of such equipment.

him by the impact of relativity on the unshakeable conviction that behind all knowledge, and yet related to all knowledge, was the external world. All measurement, whether of dead or living, of material or spiritual things, gave but symbolic knowledge – knowledge of the *structure* of that world. What could other knowledge be but knowledge of its *nature?* He could not deny that biologists had discovered something, and he was unwilling to believe that religious experience had no 'real' origin, so the biologist and the theologian had alike to be providing information about the one objective world which was hidden from the metrical physicist.

We may pause once more to see how unnecessary is this denial of the common character of all science which appears to most of us to be an obvious fact. If we accept the simple, direct statement of the scientific problem – that it is the rational correlation of experience – then physics and biology differ only in the kinds of experience with which they are concerned. The physicist correlates the artificial experiences which he produces in the laboratory, which are represented by numbers or 'pointer readings', and the biologist certain of those less artificial ones which cannot be so represented. The one creates what may be called the physical world, consisting at the present time of fundamental particles, wave functions, etc., and the other what may be called-though it less often is – the biological world, containing evolutionary development, heredity and the like. The two worlds are distinct at present, but may be expected, with the continual modification that progress always brings about in scientific conceptions, to merge ultimately into a common world, which will, however, not be a primary entity revealing or concealing its character through experience, but a rational expression of the relatedness of experience, experience alone being the primary entity.

Eddington could not bring himself to take this simple view because of his obsession with the external world. It is true that in later life he grudgingly admitted that he had done so with respect to physics. 'I accept the statement' [that science is the rational correlation of experience], he wrote in 1939, 'provided that "science" is understood to mean "physics"'.* It has taken me nearly twenty years to accept it; but by steady mastication during that period I have managed to swallow it all down bit by bit.' But he hadn't, and in fact he couldn't without vitiating much of the book which he was then writing. The very words he uses show the unreality of the admission. The statement is not one that can be swallowed bit by bit; you swallow it whole or not at all, because it does not consist of parts. What I think he meant was that he had been trying for twenty years to find reasons for rejecting it, and, having failed, had abandoned the effort.

Once admit, however, that the biologist, the psychologist and the theologian obtain a direct insight into the nature of the external world which the physicist apprehends only symbolically, and a limitless field of uncontrolled speculation is open for you to sport in. That is the danger of words and phrases representing entities which are unknowable; they invite you to attach to them ideas difficult otherwise to fit into your scheme of things, and to believe that in so doing you have done something profound. Satan finds some mischief still for idle words to do. Something has to be said about 'life, consciousness, spirit', because they do stand, clearly or obscurely, for something in our experience that is not studied by physics. The 'nature of the external world' is a phrase waiting for a content, and these are entities waiting to be placed in the external world. What could be simpler than to bring them together and to regard the external world as having the nature of 'life, consciousness, spirit'? The temptation was irresistible, and Eddington fell.

There is an example in history that should have shown him the red light. There are certain resemblances, as Eddington himself noted, between his philosophy and that of Kant. In particular, just as Eddington placed behind the physical world the inscrutable external world, so Kant placed behind all appearances the inscrutable 'thing in itself'. He was clear-sighted enough to know that he could not scrutinize the inscrutable, but his successors were not.

---

* *The Philosophy of Physical Science*, p. 185.

Fichte was the first to think he could do the impossible, and he, in effect, identified the 'thing in itself' with the ego; the unknowable something behind the things observed was that which observed. Furthermore, he foisted this monstrous notion on Kant himself. Kant, then an old man, was not long and was not uncertain in his repudiation of it, whereupon Fichte, undismayed, took on the idea for himself; if Kant would not acknowledge this philosophy, he would claim it as his own. The subsequent history of German idealism should have been an awful warning to a scientist, of all people, not to leave unknowables lying about.

Eddington became his own Fichte. The mysterious nature of the external world he identified with consciousness, which was perilously like the ego. But there was a difference, for he was not prepared to say that it was consciousness that was represented symbolically by the entities of the physical world. Instead he divided the external world into two parts, one called 'subjective' and the other 'objective'. It was the objective part only that was of the nature of 'life, consciousness, spirit', and of this we had direct knowledge through non-metrical investigation. It had 'no reason to conform to the pattern of systematization that distinguishes present-day physics'. The nature of the subjective part remained inscrutable, for physical investigation, which was our only way of approaching it, gave us only its structure. If Eddington's philosophy were in danger of becoming as influential as Kant's, it would be safe to say that before long this distinction would be eliminated, and the physical world would be hailed as being spiritual in nature. Every physical concept would be paired with the biological or psychological concept which it symbolized. The electron would perhaps become the structure of the gene and the wave function that of the id. Fortunately, this is unlikely to happen, but the notion as it stands is sufficiently bizarre. Why one part of the external world should expose itself freely to our inspection while another modestly hides behind an impenetrable symbol is a problem even worse than that which the philosophy is intended to solve. If this is the end of our pilgrimage we might as well not have started.

I come now to that aspect of Eddington's philosophy which has probably caused more misgiving than any other and which he seems to have regarded as the culminating point of the whole thing; I mean the conclusion that the laws of physics are derivable by pure reason. This, I think, was first expressed in its completed form at the end of *The Relativity Theory of Protons and Electrons*, where he says:* 'Unless the structure of the nucleus has a surprise in store for us, the conclusion seems plain – there is nothing in the whole system of laws of physics that cannot be deduced unambiguously from epistemological considerations. An intelligence, unacquainted with our universe, but acquainted with the system of thought by which the human mind interprets to itself the content of its sensory experience, should be able to attain all the knowledge of physics that we have attained by experiment. He would not deduce the particular events and objects of our experience, but he would deduce the generalizations we have based on them. For example, he would infer the existence and properties of radium, but not the dimensions of the earth.'

Let us note at the beginning that he distinguishes sharply between the *laws* of physics and the *actual entities* among which we find ourselves and which obey those laws. This is important, because he has sometimes been unjustly charged with supposing that the whole of our experience could have been predicted by a perfect reasoner, whereas in fact he supposed that none of it could.† The

---

* *The Relativity Theory of Protons and Electrons* (1936), p. 327.

† I think I should here add a comment on modem criticism which is particularly apposite to the present subject. While it is the duty of every writer to express his ideas as clearly as he can and to ensure that each sentence is literally as well as in probable effect accurate – even, where possible, when it is removed from its context – it is no less the duty of every reader to try to ascertain the ideas which the author is expressing and to refrain from analysing the defects of particular sentences as though that constituted a refutation of the author's thesis. The superior writer knows so well what he wants to say that he is particularly apt to overlook possible misinterpretations of his

laws of physics characterize the behaviour of *any conceivable* physical world, and therefore tell you nothing at all about which of the conceivable ones is the actual one; for knowledge of that we must depend on experience. This he maintained from his earliest* to his latest† writings on this subject, and it is certainly not his fault that it is not fully understood. But his actual claim, that the laws of physics – which he freely admitted had in fact been derived by generalization from the facts of experience – could have been foreseen and are inherent in the procedure by which we generalize, is sufficiently startling and demands the closest examination before judgement is passed on it. This it has not always received.

In the first place we must note that we are now concerned entirely with the physical world and not at all with the external world that lies behind it. (I have already alluded to Eddington's lapses from consistency in the use of terms, and there are occasional passages in his writings that seem to bring the external world into the matter. I am satisfied, however, that that was not his intention.) He

---

remarks by those who are dependent on those remarks for their knowledge of his meaning, and it is therefore necessary, when a reader finds an inconsistency between two separated passages, that he should make the effort to form a balanced judgement of the author's intention and criticize that rather than the faulty expression of it. This, as I say, is particularly important when reading Eddington because, as it is the main purpose of this lecture to substantiate, his unfortunate point of view made exceedingly difficult and complicated what to most of us is fairly straightforward (see *The Observatory*, 63, pp. 20-21 (1940)), and he was therefore exceptionally liable to make misleading statements. We owe it to him, and to ourselves if we wish to benefit from his unusual insight, to take his work as a whole, in its historical development, and to arrive at the essence of his contribution to thought. I do not claim to have been myself guiltless in this respect.

* *Science, Religion and Reality*, p. 210.

† *The Philosophy of Physical Science*, p. 217. A passage in his posthumous work, *Fundamental Theory* (1946), p. 31, may seem to contradict this, but careful reading shows, I think, that it does not do so.

is therefore not claiming that we could know anything about the external world by the unaided reason. The objection may be raised that if the physical world is (or has) the structure of the external world, then precognition of it must be at any rate partial precognition of the external world, for even structure is an essential characteristic. That is true, but here we must make a further analysis and distinguish the physical world, which is the structure of the external world, from *our idea* of the physical world at some particular time during the development of physics. What Eddington is talking about in this connexion is *our present knowledge* of the physical world, and he makes it clear that he does not assume that that knowledge is necessarily true.* It will make for clarity to call this subject of discussion the *apparent* physical world; it is our present idea of the true character of the physical world, the system of laws of nature which at present constitute the most generalized achievements of theoretical physics. It must be taken as a whole, and not in separated parts such as the law of gravitation, the laws of the electromagnetic field, etc.† Eddington claimed to have harmonized the two great divisions of present-day physics – relativity and quantum theory – sufficiently to enable us to regard physics as a unified subject, and it is the present formulation of the connected system of laws covering that subject that is the apparent physical universe which he claimed could be constructed by pure reason.

He not only claimed that it could be; he claimed that he had so constructed it. To analyse this claim would be, of course, to follow through the abstruse mathematical calculations that form the substance of his books, *The Relativity Theory of Protons and Electrons* and *Fundamental Theory.* This would be out of the question here, even if I were able to do it. What we can do, however, is to try to understand fully what it is that he believed he had done, and possibly to pass judgement on it from very general considerations. We note, then, that he regarded the laws of physics, by virtue of

* *The Philosophy of Physical Science*, pp. 2–3.

† See *Nature, Lond.*, 148, p. 342 (1941)

their rational origin, as being compulsory, universal and exact,* in contrast to the contingent, partial and approximate quality which must characterize laws having only an empirical justification. At the same time, however, he admitted that the final Court of Appeal with regard to all the conclusions of physics was observation and that 'every item of physical knowledge is of a form which might be submitted to the Court. It must be such that we can specify (although it may be impracticable to carry out) an observational procedure which would decide whether it is true or not. Clearly a statement cannot be tested by observation unless it is an assertion about the results of observation. *Every item of physical knowledge must therefore be an assertion of what has been or would be the result of carrying out a specified observational procedure.*'† Taken together, these claims mean that every physical observation must accord with the requirements of the rationally established laws, so that if the laws prescribe that Jupiter shall be in such and such a position at six o'clock tomorrow evening, Jupiter must be there at that time.

But now an objection immediately suggests itself. Suppose we look in that direction at six o'clock and do not see Jupiter there. Then we have submitted the case to the final Court of Appeal – observation – and it has decided against the law. How, then, can the law be compulsory? There are two answers to this. In the first place, although the law is exact, what it predicts is not certainty but *probability* of observation; it says that the probability that a specified event will happen must be exactly so and so, and if the event does not happen, the law is not thereby violated. But in such a case as this of Jupiter, the probability would be so exceedingly great that a departure from the expected place would be assigned not to this cause but to another, which we may express as follows. The laws of physics, being derived by pure reason, relate only to entities which have been postulated, not to those which have been observed. Consequently, our supposition, 'if the laws

---

* *Philosophy of Physical Science*, p. 45.

† Ibid. pp. 9–10. Eddington's italics.

prescribe that Jupiter shall be in such and such a position at six o'clock tomorrow evening', is a false one. The laws cannot possibly prescribe anything about Jupiter if by Jupiter we mean the body which astronomers observe and call by that name, because they refer only to the behaviour of entities which might exist but do not necessarily do so. When we apply them to the objects we observe we must ourselves identify each such object with one of the possible entities to which the laws relate. It is only when the identification is made correctly that the observed object must behave in the prescribed way. If, then, Jupiter does not follow the prediction, we must ascribe its failure to a false identification. This need not be simply a false identification of Jupiter with a particular postulated entity. Any false identification or lack of identification between postulated and observed entities in the whole observable universe might account for the discrepancy. In this particular case, the probable explanation would be that, to use ordinary language, an undiscovered body somewhere in the Solar System would be disturbing Jupiter, or, to use Eddington's language, a postulated possible particle at some other point in space-time had been supposed to have no corresponding observable object, whereas, in fact, it had one. Taking this into account, observation would be found to confirm the law.

It is clear that, startling as the claim is at first sight, it is not at all easy to show it to be false. We have, in fact, an actual example in the history of science of this kind of occurrence. The planet Uranus appeared to be violating Newton's law of gravitation, but the existence of another planet, Neptune, was assumed and later verified by observation. The law stood up to the test in exactly the way supposed by Eddington. But we have another instance of a planet – Mercury – which also departed from its expected path, and in that case it was the law that was abandoned, Newton's law of gravitation giving place to Einstein's. That was at a stage when physical law had not reached the generality now attained. Newton's law was incapable of being unified with the other laws of physics, such as those of electromagnetic theory and of optics, and

Eddington would not have claimed that it was epistemological, i.e. a necessary consequence of our way of approaching our problems. What his thesis amounts to is that, in the present advanced state of knowledge, any departure from expectation that you care to imagine might take place, but the explanation of the departure will then be found to be not that the law which created the expectation was wrong but that the system of objects obeying the law was wrongly selected.

The paradoxical conclusion is therefore reached that Eddington's assertion concerning the epistemological character and inviolability of the present scheme of physical law is not that experience is controlled by reason but rather the reverse – that it is entirely independent of reason. Any imaginable event may occur without breaking the laws because the possibilities allowed by the laws are so wide that any imaginable occurrence must fall within them. An analogy might make the point clearer. Suppose you are presented with a large number of different entities of widely varying character, and you want to find the most general laws that they exhibit. After some research you may discover that they can all be grouped under five different headings, namely, solids, liquids, gases, metrical concepts such as velocities and separated durations of time, and non-metrical concepts such as ideas and jokes. This could not have been foreseen; it required observation to bring it to light. You proceed further, and find a still more general fact. The first three of these groups can be classed together as gravitating things, and the last two as non-gravitating things, so that you can say that there are only two fundamentally distinct classes, which you may call 'material' and 'mental'. This again could not have been foreseen or discovered in any way but by examination of the entities. But now finally you realize that there is something that is true of all of them, without exception; they are enumerable and obey the laws of simple arithmetic. If you count out any number of them, and add any other arbitrarily chosen set, you find that the total number of entities you obtain will be altogether independent of the particular individuals you happen to select. At last you have reached a perfectly general law. But the penalty you pay for this success is that you sacrifice all knowledge of the entities with which you were presented. The laws of arithmetic are derivable epistemologically, and could have been arrived at before you began to study the data before you. No matter what those data may have been, they would still have obeyed the laws of arithmetic, so that your perfectly general law is quite unaffected by anything that anyone browsing among the entities might discover. The law was in fact discovered by experience, but it could have been discovered by someone ignorant of the fact that the set of entities existed.

That is the situation in physics as Eddington saw it. We must not forget his proviso that the nucleus may have a surprise in store for us, but, assuming without prejudice that it has not, he regarded the present scheme of physical law as of the same character as the final generalization in our analogy.

Two questions arise out of all this. First, supposing all Eddington's calculations to be valid, has he in fact shown that the fundamental laws of physics are logical necessities which could have been reached without any experience of the world from whose characteristics they have actually been deduced? The answer, I think, is unquestionably, no. In my example I described the laws of arithmetic as epistemological laws, meaning by that, as Eddington did, that they are derivable by pure reason from certain postulates. So long as those postulates are not logically incompatible with one another they may be freely chosen, and ordinary arithmetic is simply the set of conclusions that must follow from one particular choice. Certain things are defined and called 'numbers', and then their relations with one another must necessarily be what they are, just as, in Euclidean geometry, certain things are defined and called 'straight lines', 'angles', etc., and then the successive propositions necessarily follow. But now, what determines the choice of the original postulates or definitions? At one time it was thought that, so to speak, they necessarily arose in our consciousness, that we could not evade them or choose any others, and that, further, anything that we found in nature – i.e. anything that we experienced – had necessarily to exemplify the conclusions drawn from them. For example, the sum of the three angles of any naturally occurring triangle had to be equal to two right angles, and the sum of 6 things and 6 things had to be equal to 12 things whatever the things might be.

We know now that that was an error. Other geometries than Euclid's, starting from different postulates and reaching different conclusions, have been constructed, and we cannot say exactly what the sum of the three angles of a material triangle will be without measuring them to see. Similarly, we have Boolean and other queer algebras in which the ordinary laws of addition and subtraction do not hold, because those algebras proceed from original postulates other than numbers, and again we cannot say without trial that any naturally occurring system of entities will obey the arithmetical laws rather than these. In the example I chose just now they happened to do so. But suppose the entities had included durations of time that were not separated. The general, universal law would then not have been found to be true. I am told, for instance, that there are realms of experience in which, as the expression goes, the sentences run concurrently. In that case, 6 months plus 6 months equals 6 months. Or again, if we had attempted to apply the law to the measures instead of the mere enumeration of velocities, we would have found it to fail. Until early in this century it was believed that, owing to the complete universality of the laws of arithmetic, we could measure speed as we liked and still apply the laws to the results. We now know, however, that if we measure it in the customary way – namely, as so many units of space covered in one unit of time – then if we add a velocity u to a body moving with velocity v, the resulting velocity is not $u + v$ but something smaller. There are numerous other examples of the same kind.

It follows, then, that our general law did, after all, tell us something about the system of entities we were presented with. We could not have predicted it, but only by actual trial could we have found that the entities obeyed it. If we had been assured that each entity could have been represented unambiguously by the number

1, then indeed we could have predicted a great deal about their mutual relations which would have been universal, compulsory and exact, but there is nothing in the nature of things or in the nature of our minds to give us that assurance. We must discover it by experience, and by experience alone.

The effect of this on Eddington's claim is of the highest importance. Let us continue to suppose that his mathematics is impeccable and that the agreements he obtained between calculated and observed values of physical constants are valid; it still does not follow that the predictions are epistemological in the strictest sense. Before his calculations began he had to adopt certain postulates, and those postulates might have been other than they were. His justification for choosing them could only have been an empirical one. And, indeed, he was well aware of this. He cannot be reproached for not pointing it out, though I think that the extent to which, having done so, he ignored it in his more revolutionary statements has led to unnecessary misconceptions. 'To the question', he wrote,* 'whether it [epistemological knowledge] can be regarded as independent of observational experience altogether, we must, I think, answer no. A person without observational experience at all, and without that indirect knowledge of observational experience which he might gain by communication with his fellows, could not possibly attach meaning to the terms in which epistemological knowledge, like other physical knowledge, is expressed; and it would be impossible to put it into any other form which would have a meaning for him. We must grant then that the deduction of a law of nature from epistemological considerations implies antecedent observational experience.'

He went further, and tried to enumerate the particular elements of the epistemological scheme which were taken from experience; he called them 'forms of thought'.† For example, there is the practice of describing the physical world as a world, i.e. of expressing

---

* *The Philosophy of Physical Science*, pp. 24–5.

† *The Philosophy of Physical Science*, Chapter VIII.

all our knowledge as knowledge of one connected system. Again, there is the concept of analysis, i.e. the conception of the whole as divisible into parts which are permanent and precisely like one another; in effect, this means that we describe the physical world in terms of fundamental particles. He made no attempt at an exhaustive enumeration, but even this brief statement is sufficient to show how heavily he drew on experience before he began his epistemological deductions. The decision to describe the physical world in terms of identical particles, for instance, is anything but necessary. So late as the end of the nineteenth century there was a prominent school of thought that advocated what was called *energetics*, which was an endeavour to describe the physical world without using the concept of atoms and in terms only of concepts of continuity. It has died out because the atomic concept has shown itself able to deal with phenomena that would otherwise appear to be intractable, but that is entirely the result of experience, and no one can say with certainty that with further experience we shall not turn again from concepts of quanta to those of continuity. Granting Eddington's achievements everything he claimed for them, therefore, it still remains true that at bottom our knowledge rests on experience, and experience alone can tell us what particular logical form describes its interrelations.

But when all that has been said, the other question must be faced: did he in fact establish his claim to have derived a comprehensive system of physical law from postulates owing nothing to experience beyond the very general 'forms of thought' which he acknowledged? This question has not yet been finally answered, yet it is of the greatest importance that it should be. It is certain that physical law as most physicists understand it has not reached the final generalization, for the great division between field and quantum conceptions has not been bridged. Eddington, however, believed that he had bridged it, and if so his contribution to physics, as distinct from the philosophy of physics, is immense. Unfortunately, the testing of his work is extremely difficult. The nature of the mathematics involved is such that very few persons are competent to criticize it, and for them the labour would be very great. I understand that there are errors in his calculations, and certainly at least one of the observationally determined quantities whose agreement with the theoretical values is essential to success has been changed since his death. The whole question is in a very unsatisfactory state, but there can be no doubt that, even when we have reduced his claims to the lowest possible content, they are so momentous that no effort should be spared to reach a definitive evaluation of them.

Returning to the less technical side of the matter, we may derive some satisfaction from the conclusion that, after all, what he is proposing is no revolution in science, but rather the relentless pursuit of the original ideal towards a culminating point. If science originates in experience and aims at constructing a logically coherent system that will express all its interrelations, it is only to be expected that the appeal to experience will become less and less prominent as the logical system grows, finally concentrating itself into a single question when the system is complete. But throughout the process the aim is the same. Galileo's view, though we may abandon the picturesque form of its expression, is still that of Eddingtonian science: 'I incline to think that Nature first made the things themselves as she best liked, and afterwards framed the reason of man capable of conceiving (though not without great pains) some part of her secrets.'

But then our old enigma arises once more: how could Eddington make so orthodox a procedure wear so heterodox a look? We cannot directly blame the external world this time, since we are wholly on this side of that phantom, but I think it is essentially the same perversity of outlook, forcing him to describe what we do in the reverse order to that which we actually take, that is responsible for the anomaly. When we examine our procedure in science and state it in the most direct way, we realize that it is simply this. We make observations – pointer-readings, since we are now concerned with physics alone – and represent them by symbols, and we find that they are related with one another in a certain way.

We then construct a logical system with postulates so chosen that their implications agree with the relations found to hold between the observations. And that is all. As an aid to progress we try to give the logical system a picturable form, calling its elements the properties of particles or waves or something else that will suggest a way of finding further relations, but that is a means of research, not a discovery, and we freely change the picture as we advance. Everything essential in physics can be described, and all its implications deduced and their significance fully evaluated, in terms of this description.

By contrast, let us now see the same process as conceived by Eddington. We begin, not with what we do but with what we imagine must be 'true'. Standing remote in the background is the awful Reality of the external world, mysterious, inaccessible. In front is the physical world which presents, but only symbolically, the structure of a part of it. Of this we can attain something called knowledge which, though its name suggests subjectivity, is in effect an objective entity that stands in front of the physical world of which it is a representation. But not a precise representation. It shows us only the probable character of the physical world, and this is what is depicted in the equations of physics. We have still to reach these equations, and before us there is a road called experience. We have for centuries been toiling painfully along this road, and at last we have reached the fundamental equations, so that now we know all that theoretical physics can tell us, namely, the probable character of the structure of part of Reality. But on looking behind we can see that there is another road called reason which also comes out at the same equations. We follow it back, and at the far end we find ourselves again in the land of pure ignorance from which we began our journey. We could therefore have reached the same goal if we had proceeded that way instead of by experience. But we see also that from this point an indefinitely large number of other roads go out. They are all marked reason, and there is nothing in the roads themselves to tell us which one leads to the same destination as the road of experience. There is, however, one way by which we can discover this. If we compare them with the road of experience we see that one, and only one, is parallel to it. If, then, we follow that, we shall get to the equations of physics and so learn the probable state of the symbolical structure of the external world without calling on experience at all. That is the final conclusion that Eddington reached.

When this amazing conception is laid bare, we can only pause before it in mute wonder. How is it possible, we eventually ask, that what is in essence so simple can be twisted into a form so intricate? I do not think an answer can be given on any supposition other than that which I have indicated, that the description had to grant full recognition to both the Victorian external world, from which all life had gone, and the necessary implications of the relativity theory. The practical difficulties of thinking in terms of this labyrinth are obvious enough, but what is of far greater moment is the essentially wrong representation which it gives of the place of experience in science. Instead of showing experience as the origin and centre of interest of the whole effort, it leads us to regard it as merely the lesser of two alternative means towards a greater end. Intent on preserving what it misconceives as the 'Truth', it has lost the 'Way' and the 'Life'. Relativity was not so much a revolution in science as a purification; it recalled physics from its traffic with metaphysical notions to its true concern with what we actually observe. By its acknowledgement of the final authority and inviolability of experience it opened the possibility, for the first time in modern history, of granting full licence to science to pursue the rational correlation of experience without danger of conflict with art, religion and other forms of philosophy so long as they also assert nothing that is not grounded in experience. In Eddington's philosophy of science that inestimable clarification is obscured. The essence of the matter is there, but instead of being illuminated it is shrouded in mist.

That is one aspect of the matter, but there is another on which it is more pleasant to dwell. It is easy enough to recognize the enormous burden that Eddington's philosophy must have imposed on its victim, but the incredible fact remains that he not only bore it but used it to reach heights which his contemporaries, for all their advantages of equipment, could not attain. It is not only that he escaped the throes of instinct at strife with reason, of which the Victorian age exhibits so many spectacles, described in Tennyson's *In Memoriam*, Huxley's Romanes Lecture, and countless other records. That struggle he evaded so easily that it is only with an effort that we become aware that it should have involved him in its toils. But the chief marvel is that he could use the dread machinery in which most of us can only become hopelessly entangled to produce a theory which, whatever the ultimate verdict on it may be, is beyond all question a work of the highest genius. It is notoriously dangerous to prophesy unless you know, and in a matter of such difficulty I am very far indeed from knowing, yet I do not hesitate to express the belief that when Eddington's fundamental theory is translated from the terms of his unspeakable philosophy into a language that ordinary mortals can understand, it will be found not only to be a work of outstanding technical skill, but also to contain scientific truths which he alone in his generation had the depth of vision to perceive. It seems a tragedy that a man with such incomparable sight should have been placed at such a point of disadvantage, but let us not fail to recognize that his description of what he saw, though abnormal and in our eyes distorted, was in all probability true. I said that he did not know what he was doing; but I believe that what he did was supremely great.

AN EMPIRICIST'S VIEW OF THE NATURE OF RELIGIOUS BELIEF

Richard B. Braithwaite

FBA

THE NINTH

ARTHUR STANLEY EDDINGTON

MEMORIAL LECTURE

22 November 1955

An Empiricist's View of the Nature of Religious Belief

'THE MEANING OF A scientific statement is to be ascertained by reference to the steps which would be taken to verify it.' Eddington wrote this in 1939. Unlike his heterodox views of the *a priori* and epistemological character of the ultimate laws of physics, this principle is in complete accord with contemporary philosophy of science; indeed it was Eddington's use of it in his expositions of relativity theory in the early 1920s that largely contributed to its becoming the orthodoxy. Eddington continued his passage by saying: 'This [principle] will be recognised as a tenet of logical positivism – only it is there extended to all statements.'* Just as the tone was set to the empiricist tradition in British philosophy – the tradition running from Locke through Berkeley, Hume, Mill to Russell in our own time – by Locke's close association with the scientific work of Boyle and the early Royal Society, so the contemporary development of empiricism popularly known as logical positivism has been greatly influenced by the revolutionary changes this century in physical theory and by the philosophy of science which physicists concerned with these changes – Einstein and Heisenberg as well as Eddington – have thought most consonant with relativity and quantum physics. It is therefore, I think, proper for me to take the verificational principle of meaning, and a natural adaptation of it, as that aspect of contemporary scientific thought whose bearing upon the philosophy of religion I shall discuss this afternoon. Eddington, in the passage from which I have quoted, applied the verificational principle to the meaning of scientific statements only. But we shall see that it will be necessary, and concordant with an empiricist way of thinking, to modify the

* A. S. Eddington, *The Philosophy of Physical Science* (1939), p. 189

principle by allowing *use* as well as *verifiability* to be a criterion for meaning; so I believe that all I shall say will be in the spirit of a remark with which Eddington concluded an article published in 1926: 'The scientist and the religious teacher may well be content to agree that the *value* of any hypothesis extends just so far as it is verified by actual experience.'*

I will start with the verificational principle in the form in which it was originally propounded by logical positivists – that the meaning of any statement is given by its method of verification.†

The implication of this general principle for the problem of religious belief is that the primary question becomes, not whether a religious statement such as that a personal God created the world is true or is false, but how it could be known either to be true or to be false. Unless this latter question can be answered, the religious statement has no ascertainable meaning and there is nothing expressed by it to be either true or false. Moreover a religious statement cannot be believed without being understood, and it can only be understood by an understanding of the circumstances which would verify or falsify it. Meaning is not logically prior to the possibility of verification: we do not first learn the meaning of a statement, and afterwards consider what would make us call it true or false; the two understandings are one and indivisible.

It would not be correct to say that discussions of religious belief before this present century have always ignored the problem of meaning, but until recently the emphasis has been upon the question of the truth or the reasonableness of religious beliefs rather than upon the logically prior question as to the meaning of the statements expressing the beliefs. The argument usually proceeded as if we all knew what was meant by the statement that a personal God created the world; the point at issue was whether or not this statement was true, or whether there were good reasons

---

* *Science, Religion and Reality*, ed. by J. Needham (1926), p. 218 (my italics)

† The principle was first explicitly stated by F. Waismann, in *Erkenntnis*, vol. 1 (1930), p. 229

for believing it. But if the meaning of a religious statement has to be found by discovering the steps which must be taken to ascertain its truth-value, an examination of the methods for testing the statement for truth-value is an essential preliminary to any discussion as to which of the truth-values – truth or falsity – holds of the statement.

There are three classes of statement whose method of truth-value testing is in general outline clear: statements about particular matters of empirical fact, scientific hypotheses and other general empirical statements, and the logically necessary statements of logic and mathematics (and their contradictories). Do religious statements fall into any of these three classes? If they do, the problem of their meaningfulness will be solved: their truth-values will be testable by the methods appropriate to empirical statements, particular or general, or to mathematical statements. It seems to me clear that religious statements, as they are normally used, have no place in this trichotomy. I shall give my reasons very briefly, since I have little to add here to what other empiricist philosophers have said.

(1) Statements about particular empirical facts are testable by direct observation. The only facts that can be directly known by observation are that the things observed have certain observable properties or stand in certain observable relations to one another. If it is maintained that the *existence* of God is known by observation, for example, in the 'self-authenticating' experience of 'meeting God', the term 'God' is being used merely as part of the description of that particular experience. Any interesting theological proposition, e.g. that God is personal, will attribute a property to God which is not an observable one and so cannot be known by direct observation. Comparison with our knowledge of other people is an unreal comparison. I can get to know things about an intimate friend at a glance, but this knowledge is not self-authenticating; it is based upon a great deal of previous knowledge about the connexion between facial and bodily expressions and states of mind.

(2) The view that would class religious statements with scientific hypotheses must be taken much more seriously. It would be very unplausible if a Baconian methodology of science had to be employed, and scientific hypotheses taken as simple generalizations from particular instances, for then there could be no understanding of a general theological proposition unless particular instances of it could be directly observed. But an advanced science has progressed far beyond its natural history stage; it makes use in its explanatory hypotheses of concepts of a high degree of abstractness and at a far remove from experience. These theoretical concepts are given a meaning by the place they occupy in a deductive system consisting of hypotheses of different degrees of generality in which the least general hypotheses, deducible from the more general ones, are generalizations of observable facts. So it is no valid criticism of the view that would treat God as an empirical concept entering into an explanatory hypothesis to say that God is not directly observable. No more is an electric field of force or a Schrödinger wave-function. There is no prima facie objection to regarding such a proposition as that there is a God who created and sustains the world as an explanatory scientific hypothesis.

But if a set of theological propositions are to be regarded as scientific explanations of facts in the empirical world, they must be refutable by experience. We must be willing to abandon them if the facts prove different from what we think they are. A hypothesis which is consistent with every possible empirical fact is not an empirical one. And though the theoretical concepts in a hypothesis need not be explicitly definable in terms of direct observation – indeed they must not be if the system is to be applicable to novel situations – yet they must be related to some and not to all of the possible facts in the world in order to have a non-vacuous significance. If there is a personal God, how would the world be different if there were not? Unless this question can be answered God's existence cannot be given an empirical meaning.

At earlier times in the history of religion God's personal existence has been treated as a scientific hypothesis subjectable to empirical test. Elijah's contest with the prophets of Baal was an experiment to test the hypothesis that Jehovah and not Baal controlled the physical world. But most educated believers at the present time do not think of God as being detectable in this sort of way, and hence do not think of theological propositions as explanations of facts in the world of nature in the way in which established scientific hypotheses are.

It may be maintained, however, that theological propositions explain facts about the world in another way. Not perhaps the physical world, for physical science has been so successful with its own explanations; but the facts of biological and psychological development. Now it is certainly the case that a great deal of traditional Christian language – phrases such as 'original sin', 'the old Adam', 'the new man', 'growth in holiness' – can be given meanings within statements expressing general hypotheses about human personality. Indeed it is hardly too much to say that almost all statements about God as immanent, as an indwelling spirit, can be interpreted as asserting psychological facts in metaphorical language. But would those interpreting religious statements in this way be prepared to abandon them if the empirical facts were found to be different? Or would they rather re-interpret them to fit the new facts? In the latter case the possibility of interpreting them to fit experience is not enough to give an empirical meaning to the statements. Mere consistency with experience without the possibility of inconsistency does not determine meaning. And a metaphorical description is not in itself an explanation. This criticism also holds against attempts to interpret theism as an explanation of the course of history, unless it is admitted (which few theists would be willing to admit) that, had the course of history been different in some specific way, God would not have existed.

Philosophers of religion who wish to make empirical facts relevant to the meaning of religious statements but at the same time desire to hold on to these statements whatever the empirical facts may be are indulging, I believe, in a sort of 'double-think' attitude: they want to hold that religious statements both are about the actual world (i.e. are empirical statements) and also are not refutable in any possible world, the characteristic of statements which are logically necessary.

(3) The view that statements of natural theology resemble the propositions of logic and mathematics in being logically necessary would have as a consequence that they make no assertion of existence. Whatever exactly be the status of logically necessary propositions, Hume and Kant have conclusively shown that they are essentially hypothetical. $2 + 3 = 5$ makes no assertion about there being any things in the world; what it says is that, *if* there is a class of five things in the world, *then* this class is the union of two mutually exclusive subclasses one comprising two and the other comprising three things. The logical-positivist thesis, due to Wittgenstein, that the truth of this hypothetical proposition is verified not by any logical fact about the world but by the way in which we use numerical symbols in our thinking goes further than Kant did in displacing logic and mathematics from the world of reality. But it is not necessary to accept this more radical thesis in order to agree with Kant that no logically necessary proposition can assert existence; and this excludes the possibility of regarding theological propositions as logically necessary in the way in which the hypothetical propositions of mathematics and logic are necessary.

The traditional arguments for a Necessary God – the ontological and the cosmological – were elaborated by Anselm and the scholastic philosophers before the concurrent and inter-related development of natural science and of mathematics had enabled necessity and contingency to be clearly distinguished. The necessity attributed by these arguments to the being of God may perhaps be different from the logical necessity of mathematical truths; but, if so, no method has been provided for testing the truth-value of the statement that God is necessary being, and consequently no way given for assigning meaning to the terms 'necessary being' and 'God'.

If religious statements cannot be held to fall into any of these three classes, their method of verification cannot be any of the standard methods applicable to statements falling in these classes. Does this imply that religious statements are not verifiable, with the corollary, according to the verificational principle, that they have no meaning and, though they purport to say something, are in fact nonsensical sentences? The earlier logical positivists thought so: they would have echoed the demand of their precursor Hume that a volume ('of divinity or school metaphysics') which contains neither 'any abstract reasoning concerning quantity or number' nor 'any experimental reasoning concerning matter of fact and existence' should be committed to the flames; though their justification for the holocaust would be even more cogent than Hume's. The volume would not contain even 'sophistry and illusion': it would contain nothing but meaningless marks of printer's ink.

Religious statements, however, are not the only statements which are unverifiable by standard methods; moral statements have the same peculiarity. A moral principle, like the utilitarian principle that a man ought to act so as to maximize happiness, does not seem to be either a logically necessary or a logically impossible proposition. But neither does it seem to be an empirical proposition, all the attempts of ethical empiricists to give naturalistic analyses having failed. Though a tough-minded logical positivist might be prepared to say that all religious statements are sound and fury, signifying nothing, he can hardly say that of all moral statements. For moral statements have a use in guiding conduct; and if they have a use they surely have a meaning – in some sense of meaning. So the verificational principle of meaning in the hands of empiricist philosophers in the 1930s became modified either by a glossing of the term 'verification' or by a change of the verification principle into the use principle: the meaning of any statement is given by the way in which it is used.*

Since I wish to continue to employ verification in the restricted sense of ascertaining truth-value, I shall take the principle of mean-

* See L. Wittgenstein, *Philosophical Investigations* (1953), especially §§ 340, 353, 559 and 560.

ing in this new form in which the word 'verification' has disappeared. But in removing this term from the statement of the principle, there is no desertion from the spirit of empiricism. The older verificational principle is subsumed under the new use principle: the use of an empirical statement derives from the fact that the statement is empirically verifiable, and the logical-positivist thesis of the 'linguistic' character of logical and mathematical statements can be equally well, if not better, expressed in terms of their use than of their method of verification. Moreover the only way of discovering how a statement is used is by an empirical enquiry; a statement need not itself be empirically verifiable, but that it is used in a particular way is always a straightforwardly empirical proposition.

The meaning of any statement, then, will be taken as being given by the way it is used. The kernel for an empiricist of the problem of the nature of religious belief is to explain, in empirical terms, how a religious statement is used by a man who asserts it in order to express his religious conviction.

Since I shall argue that the primary element in this use is that the religious assertion is used as a moral assertion, I must first consider how moral assertions are used. According to the view developed by various moral philosophers since the impossibility of regarding moral statements as verifiable propositions was recognized, a moral assertion is used to express an *attitude* of the man making the assertion. It is not used to assert the proposition that he has the attitude – a verifiable psychological proposition; it is used to show forth or evince his attitude. The attitude is concerned with the action which he asserts to be right or to be his duty, or the state of affairs which he asserts to be good; it is a highly complex state, and contains elements to which various degrees of importance have been attached by moral philosophers who have tried to work out an 'ethics without propositions'. One element in the attitude is a feeling of approval towards the action; this element was taken as the fundamental one in the first attempts, and views of ethics without propositions are frequently lumped together as 'emotive' theories of ethics. But discussion of the subject during the last twenty years has made it clear, I think, that no emotion or feeling of approval is fundamental to the use of moral assertions; it may be the case that the moral asserter has some specific feeling directed on to the course of action said to be right, but this is not the most important element in his 'pro-attitude' towards the course of action: what is primary is his intention to perform the action when the occasion for it arises.

The form of ethics without propositions which I shall adopt is therefore a conative rather than an emotive theory: it makes the primary use of a moral assertion that of expressing the intention of the asserter to act in a particular sort of way specified in the assertion. A utilitarian, for example, in asserting that he ought to act so as to maximize happiness, is thereby declaring his intention to act, to the best of his ability, in accordance with the policy of utilitarianism: he is not asserting any proposition, or necessarily evincing any feeling of approval; he is subscribing to a policy of action. There will doubtless be empirical propositions which he may give as reasons for his adherence to the policy (e.g. that happiness is what all, or what most people, desire), and his having the intention will include his understanding what is meant by pursuing the policy, another empirically verifiable proposition. But there will be no specifically moral proposition which he will be asserting when he declares his intention to pursue the policy. This account is fully in accord with the spirit of empiricism, for whether or not a man has the intention of pursuing a particular behaviour policy can be empirically tested, both by observing what he does and by hearing what he replies when he is questioned about his intentions.

Not all expressions of intentions will be moral assertions: for the notion of morality to be applicable it is necessary either that the policy of action intended by the asserter should be a general policy (e.g. the policy of utilitarianism) or that it should be subsumable under a general policy which the asserter intends to follow and which he would give as the reason for his more specific intention. There are difficulties and vaguenesses in the notion of a general policy of action, but these need not concern us here. All that we require is that, when a man asserts that he ought to do so-and-so, he is using the assertion to declare that he resolves, to the best of his ability, to do so-and-so. And he will not necessarily be insincere in his assertion if he suspects, at the time of making it, that he will not have the strength of character to carry out his resolution.

The advantage this account of moral assertions has over all others, emotive non-propositional ones as well as cognitive propositional ones, is that it alone enables a satisfactory answer to be given to the question: What is the reason for my doing what I think I ought to do? The answer it gives is that, since my thinking that I ought to do the action is my intention to do it if possible, the reason why I do the action is simply that I intend to do it, if possible. On every other ethical view there will be a mysterious gap to be filled somehow between the moral judgment and the intention to act in accordance with it: there is no such gap if the primary use of a moral assertion is to declare such an intention.

Let us now consider what light this way of regarding moral assertions throws upon assertions of religious conviction. The idealist philosopher McTaggart described religion as 'an emotion resting on a conviction of a harmony between ourselves and the universe at large',* and many educated people at the present time would agree with him. If religion is essentially concerned with emotion, it is natural to explain the use of religious assertions on the lines of the original emotive theory of ethics and to regard them as primarily evincing religious feelings or emotions. The assertion, for example, that God is our Heavenly Father will be taken to express the asserter's feeling secure in the same way as he would feel secure in his father's presence. But explanations of religion in terms of feeling, and of religious assertions as expressions of such feelings, are usually propounded by people who stand outside any religious system; they rarely satisfy those who speak from inside. Few religious men would be prepared to admit that their religion

* J. M. E. McTaggart, *Some Dogmas of Religion* (1906), p. 3.

was a matter merely of feeling: feelings – of joy, of consolation, of being at one with the universe – may enter into their religion, but to evince such feelings is certainly not the primary use of their religious assertions.

This objection, however, does not seem to me to apply to treating religious assertions in the conative way in which recent moral philosophers have treated moral statements – as being primarily declarations of adherence to a policy of action, declarations of commitment to a way of life. That the way of life led by the believer is highly relevant to the sincerity of his religious conviction has been insisted upon by all the moral religions, above all, perhaps, by Christianity. 'By their fruits ye shall know them.' The view which I put forward for your consideration is that the intention of a Christian to follow a Christian way of life is not only the criterion for the sincerity of his belief in the assertions of Christianity; it is the criterion for the meaningfulness of his assertions. Just as the meaning of a moral assertion is given by its use in expressing the asserter's intention to act, so far as in him lies, in accordance with the moral principle involved, so the meaning of a religious assertion is given by its use in expressing the asserter's intention to follow a specified policy of behaviour. To say that it is belief in the dogmas of religion which is the cause of the believer's intending to behave as he does is to put the cart before the horse: it is the intention to behave which constitutes what is known as religious conviction.

But this assimilation of religious to moral assertions lays itself open to an immediate objection. When a moral assertion is taken as declaring the intention of following a policy, the form of the assertion itself makes it clear what the policy is with which the assertion is concerned. For a man to assert that a certain policy ought to be pursued, which on this view is for him to declare his intention of pursuing the policy, presupposes his understanding what it would be like for him to pursue the policy in question. I cannot resolve not to tell a lie without knowing what a lie is. But if a religious assertion is the declaration of an intention to carry out a certain policy, what policy does it specify? The religious statement itself will not explicitly refer to a policy, as does a moral statement; how then can the asserter of the statement know what is the policy concerned, and how can he intend to carry out a policy if he does not know what the policy is? I cannot intend to do something I know not what.

The reply to this criticism is that, if a religious assertion is regarded as representative of a large number of assertions of the same religious system, the body of assertions of which the particular one is a representative specimen is taken by the asserter as implicitly specifying a particular way of life. It is no more necessary for an empiricist philosopher to explain the use of a religious statement taken in isolation from other religious statements than it is for him to give a meaning to a scientific hypothesis in isolation from other scientific hypotheses. We understand scientific hypotheses, and the terms that occur in them, by virtue of the relation of the whole system of hypotheses to empirically observable facts; and it is the whole system of hypotheses, not one hypothesis in isolation, that is tested for its truth-value against experience. So there are good precedents, in the empiricist way of thinking, for considering a system of religious assertions as a whole, and for examining the way in which the whole system is used.

If we do this the fact that a system of religious assertions has a moral function can hardly be denied. For to deny it would require any passage from the assertion of a religious system to a policy of action to be mediated by a moral assertion. I cannot pass from asserting a fact, of whatever sort, to intending to perform an action, without having the hypothetical intention to intend to do the action if I assert the fact. This holds however widely fact is understood – whether as an empirical fact or as a non-empirical fact about goodness or reality. Just as the intention-to-act view of moral assertions is the only view that requires no reason for my doing what I assert to be my duty, so the similar view of religious assertions is the only one which connects them to ways of life without requiring an additional premiss. Unless a Christian's assertion that God is love (*agape*) – which I take to epitomize the assertions of the Christian religion – be taken to declare his intention to follow an agapeistic way of life, he could be asked what is the connexion between the assertion and the intention, between Christian belief and Christian practice. And this question can always be asked if religious assertions are separated from conduct. Unless religious principles are moral principles, it makes no sense to speak of putting them into practice.

The way to find out what are the intentions embodied in a set of religious assertions, and hence what is the meaning of the assertions, is by discovering what principles of conduct the asserter takes the assertions to involve. These may be ascertained both by asking him questions and by seeing how he behaves, each test being supplemental to the other. If what is wanted is not the meaning of the religious assertions made by a particular man but what the set of assertions would mean were they to be made by anyone of the same religion (which I will call their *typical* meaning), all that can be done is to specify the form of behaviour which is in accordance with what one takes to be the fundamental moral principles of the religion in question. Since different people will take different views as to what these fundamental moral principles are, the typical meaning of religious assertions will be different for different people. I myself take the typical meaning of the body of Christian assertions as being given by their proclaiming intentions to follow an agapeistic way of life, and for a description of this way of life – a description in general and metaphorical terms, but an empirical description nevertheless – should quote most of the Thirteenth Chapter of I Corinthians. Others may think that the Christian way of life should be described somewhat differently, and will therefore take the typical meaning of the assertions of Christianity to correspond to their different view of its fundamental moral teaching.

My contention then is that the primary use of religious assertions is to announce allegiance to a set of moral principles: without such allegiance there is no 'true religion'. This is borne out by all the accounts of what happens when an unbeliever becomes converted to a religion. The conversion is not only a change in the propositions believed – indeed there may be no specifically intellectual change at all; it is a change in the state of will. An excellent instance is C. S. Lewis's recently published account of his conversion from an idealist metaphysic – 'a religion [as he says] that cost nothing' – to a theism where he faced (and he quotes George MacDonald's phrase) 'something to be neither more nor less nor other than *done*'. There was no intellectual change, for (as he says) 'there had long been an ethic (theoretically) attached to my Idealism': it was the recognition that he had to do something about it, that 'an attempt at complete virtue must be made'.* His conversion was a re-orientation of the will.

In assimilating religious assertions to moral assertions I do not wish to deny that there are any important differences. One is the fact already noticed that usually the behaviour policy intended is not specified by one religious assertion in isolation. Another difference is that the fundamental moral teaching of the religion is frequently given, not in abstract terms, but by means of concrete examples – of how to behave, for instance, if one meets a man set upon by thieves on the road to Jericho. A resolution to behave like the good Samaritan does not, in itself, specify the behaviour to be resolved upon in quite different circumstances. However, absence of explicitly recognized general principles does not prevent a man from acting in accordance with such principles; it only makes it more difficult for a questioner to discover upon what principles he is acting. And the difficulty is not only one way round. If moral principles are stated in the most general form, as most moral philosophers have wished to state them, they tend to become so far removed from particular courses of conduct that it is difficult, if not impossible, to give them any precise content. It may be hard to find out what exactly is involved in the imitation of Christ; but it is not very easy to discover what exactly is meant by the pursuit

---

* C. S. Lewis, *Surprised by Joy* (1955), pp. 198, 212–13.

of Aristotle's *eudaemonia* or of Mill's *happiness*. The tests for what it is to live agapeistically are as empirical as are those for living in quest of happiness; but in each case the tests can best be expounded in terms of examples of particular situations.

A more important difference between religious and purely moral principles is that, in the higher religions at least, the conduct preached by the religion concerns not only external but also internal behaviour. The conversion involved in accepting a religion is a conversion, not only of the will, but of the heart. Christianity requires not only that you should behave towards your neighbour as if you loved him as yourself: it requires that you should love him as yourself. And though I have no doubt that the Christian concept of *agape* refers partly to external behaviour – the agapeistic behaviour for which there are external criteria – yet being filled with *agape* includes more than behaving agapeistically externally: it also includes an agapeistic frame of mind. I have said that I cannot regard the expression of a feeling of any sort as the primary element in religious assertion; but this does not imply that intention to feel in a certain way is not a primary element, nor that it cannot be used to discriminate religious declarations of policy from declarations which are merely moral. Those who say that Confucianism is a code of morals and not, properly speaking, a religion are, I think, making this discrimination.

The resolution proclaimed by a religious assertion may then be taken as referring to inner life as well as to outward conduct. And the superiority of religious conviction over the mere adoption of a moral code in securing conformity to the code arises from a religious conviction changing what the religious man wants. It may be hard enough to love your enemy, but once you have succeeded in doing so it is easy to behave lovingly towards him. But if you continue to hate him, it requires a heroic perseverance continually to behave as if you loved him. Resolutions to feel, even if they are only partly fulfilled, are powerful reinforcements of resolutions to act.

But though these qualifications may be adequate for distinguishing religious assertions from purely moral ones, they are not sufficient to discriminate between assertions belonging to one religious system and those belonging to another system in the case in which the behaviour policies, both of inner life and of outward conduct, inculcated by the two systems are identical. For instance, I have said that I take the fundamental moral teaching of Christianity to be the preaching of an agapeistic way of life. But a Jew or a Buddhist may, with considerable plausibility, maintain that the fundamental moral teaching of his religion is to recommend exactly the same way of life. How then can religious assertions be distinguished into those which are Christian, those which are Jewish, those which are Buddhist, by the policies of life which they respectively recommend if, on examination, these policies turn out to be the same?

Many Christians will, no doubt, behave in a specifically Christian manner in that they will follow ritual practices which are Christian and neither Jewish nor Buddhist. But though following certain practices may well be the proper test for membership of a particular religious society, a church, not even the most ecclesiastically-minded Christian will regard participation in a ritual as the fundamental characteristic of a Christian way of life. There must be some more important difference between an agapeistically policied Christian and an agapeistically policied Jew than that the former attends a church and the latter a synagogue.

The really important difference, I think, is to be found in the fact that the intentions to pursue the behaviour policies, which may be the same for different religions, are associated with thinking of different *stories* (or sets of stories). By a story I shall here mean a proposition or set of propositions which are straightforwardly empirical propositions capable of empirical test and which are thought of by the religious man in connexion with his resolution to follow the way of life advocated by his religion. On the assumption that the ways of life advocated by Christianity and by Buddhism are essentially the same, it will be the fact that the intention to follow this way of life is associated in the mind of a Christian with thinking of one set of stories (the Christian stories) while it is associated in the mind of a Buddhist with thinking of another set of stories (the Buddhist stories) which enables a Christian assertion to be distinguished from a Buddhist one.

A religious assertion will, therefore, have a propositional element which is lacking in a purely moral assertion, in that it will refer to a story as well as to an intention. The reference to the story is not an assertion of the story taken as a matter of empirical fact: it is a telling of the story, or an alluding to the story, in the way in which one can tell, or allude to, the story of a novel with which one is acquainted. To assert the whole set of assertions of the Christian religion is both to tell the Christian doctrinal story and to confess allegiance to the Christian way of life.

The story, I have said, is a set of empirical propositions, and the language expressing the story is given a meaning by the standard method of understanding how the story-statements can be verified. The empirical story-statements will vary from Christian to Christian; the doctrines of Christianity are capable of different empirical interpretations, and Christians will differ in the interpretations they put upon the doctrines. But the interpretations will all be in terms of empirical propositions. Take, for example, the doctrine of Justification by means of the Atonement. Matthew Arnold imagined it in terms of

> ... a sort of infinitely magnified and improved Lord Shaftesbury, with a race of vile offenders to deal with, whom his natural goodness would incline him to let off, only his sense of justice will not allow it; then a younger Lord Shaftesbury, on the scale of his father and very dear to him, who might live in grandeur and splendour if he liked, but who prefers to leave his home, to go and live among the race of offenders, and to be put to an ignominious death, on condition that his merits shall be counted against their demerits, and that his father's goodness shall be restrained no longer from taking effect, but any offender shall be admitted to the benefit of it on simply pleading the satisfaction

made by the son; – and then, finally, a third Lord Shaftesbury, still on the same high scale, who keeps very much in the background, and works in a very occult manner, but very efficaciously nevertheless, and who is busy in applying everywhere the benefits of the son's satisfaction and the father's goodness.*

Arnold's 'parable of the three Lord Shaftesburys' got him into a lot of trouble: he was 'indignantly censured' (as he says) for wounding 'the feelings of the religious community by turning into ridicule an august doctrine, the object of their solemn faith'.† But there is no other account of the Anselmian doctrine of the Atonement that I have read which puts it in so morally favourable a light. Be that as it may, the only way in which the doctrine can be understood verificationally is in terms of human beings – mythological beings, it may be, who never existed, but who nevertheless would have been empirically observable had they existed.

For it is not necessary, on my view, for the asserter of a religious assertion to believe in the truth of the story involved in the assertions: what is necessary is that the story should be entertained in thought, i.e. that the statement of the story should be understood as having a meaning. I have secured this by requiring that the story should consist of empirical propositions. Educated Christians of the present day who attach importance to the doctrine of the Atonement certainly do not believe an empirically testable story in Matthew Arnold's or any other form. But it is the fact that entertainment in thought of this and other Christian stories forms the context in which Christian resolutions are made which serves to distinguish Christian assertions from those made by adherents of another religion, or of no religion.

What I am calling a *story* Matthew Arnold called a *parable* and a *fairy-tale*. Other terms which might be used are *allegory*, *fable*, *tale*, *myth*. I have chosen the word 'story' as being the most neutral

---

* Matthew Arnold, *Literature and Dogma* (1873), pp. 306–7.

† Matthew Arnold, *God and the Bible* (1875), pp. 18–19.

term, implying neither that the story is believed nor that it is disbelieved. The Christian stories include straightforward historical statements about the life and death of Jesus of Nazareth; a Christian (unless he accepts the unplausible Christ-myth theory) will naturally believe some or all of these. Stories about the beginning of the world and of the Last Judgment as facts of past or of future history are believed by many unsophisticated Christians. But my contention is that belief in the truth of the Christian stories is not the proper criterion for deciding whether or not an assertion is a Christian one. A man is not, I think, a professing Christian unless he both proposes to live according to Christian moral principles and associates his intention with thinking of Christian stories; but he need not believe that the empirical propositions presented by the stories correspond to empirical fact.

But if the religious stories need not be believed, what function do they fulfil in the complex state of mind and behaviour known as having a religious belief? How is entertaining the story related to resolving to pursue a certain way of life? My answer is that the relation is a psychological and causal one. It is an empirical psychological fact that many people find it easier to resolve upon and to carry through a course of action which is contrary to their natural inclinations if this policy is associated in their minds with certain stories. And in many people the psychological link is not appreciably weakened by the fact that the story associated with the behaviour policy is not believed. Next to the Bible and the Prayer Book the most influential work in English Christian religious life has been a book whose stories are frankly recognized as fictitious – Bunyan's *Pilgrim's Progress*; and some of the most influential works in setting the moral tone of my generation were the novels of Dostoevsky. It is completely untrue, as a matter of psychological fact, to think that the only intellectual considerations which affect action are beliefs: it is *all* the thoughts of a man that determine his behaviour; and these include his phantasies, imaginations, ideas of what he would wish to be and do, as well as the propositions which he believes to be true.

This important psychological fact, a commonplace to all students of the influence of literature upon life, has not been given sufficient weight by theologians and philosophers of religion. It has not been altogether ignored; for instance, the report of the official Commission on Doctrine in the Church of England, published in 1938, in a section entitled 'On the application to the Creeds of the conception of symbolic truth' says: 'Statements affirming particular facts may be found to have value as pictorial expressions of spiritual truths, even though the supposed facts themselves did not actually happen ... It is not therefore of necessity illegitimate to accept and affirm particular clauses of the Creeds while understanding them in this symbolic sense.'* But the patron saint whom I claim for my way of thinking is that great but neglected Christian thinker Matthew Arnold, whose parable of the three Lord Shaftesburys is a perfect example of what I take a religious story to be. Arnold's philosophy of religion has suffered from his striking remarks being lifted from their context: his description of religion as *morality touched by emotion* does not adequately express his view of the part played by imagination in religion. Arnold's main purpose in his religious writings was that of 'cementing the alliance between the imagination and conduct' † by regarding the propositional element in Christianity as 'literature' rather than as 'dogma'. Arnold was not prepared to carry through his programme completely; he regarded *the Eternal not ourselves that makes for righteousness* more dogmatically than fictionally. But his keen insight into the imaginative and poetic element in religious belief as well as his insistence that religion is primarily concerned with guiding conduct make him a profound philosopher of religion as well as a Christian teacher full of the 'sweet reasonableness' he attributed to Christ.

God's wisdom and God's goodness! – Ay, but fools
Mis-define these till God knows them no more.

---

* *Doctrine in the Church of England* (1938), pp. 37–8.

† Matthew Arnold, *God and the Bible* (1875), p. xiii.

Wisdom and goodness, they are God! – what schools
Have yet so much as heard this simpler lore?*

To return to our philosophizing. My contention that the propositional element in religious assertions consists of stories interpreted as straightforwardly empirical propositions which are not, generally speaking, believed to be true has the great advantage of imposing no restriction whatever upon the empirical interpretation which can be put upon the stories. The religious man may interpret the stories in the way which assists him best in carrying out the behaviour policies of his religion. He can, for example, think of the three persons of the Trinity in visual terms, as did the great Christian painters, or as talking to one another, as in the poems of St John of the Cross. And since he need not believe the stories he can interpret them in ways which are not consistent with one another. It is disastrous for anyone to try to believe empirical propositions which are mutually inconsistent, for the courses of action appropriate to inconsistent beliefs are not compatible. The needs of practical life require that the body of believed propositions should be purged of inconsistency. But there is no action which is appropriate to thinking of a proposition without believing it; thinking of it may, as I have said, produce a state of mind in which it is easier to carry out a particular course of action, but the connexion is causal: there is no intrinsic connexion between the thought and the action. Indeed a story may provide better support for a long range policy of action if it contains inconsistencies. The Christian set of stories, for example, contains both a pantheistic sub-set of stories in which everything is a part of God and a dualistic Manichaean sub-set of stories well represented by St Ignatius Loyola's allegory of a conflict between the forces of righteousness under the banner of Christ and the forces of darkness under Lucifer's banner. And the Marxist religion's set of stories contains both stories about an inevitable perfect society and stories about a class

* From Matthew Arnold's sonnet 'The Divinity' (1867).

war. In the case of both religions the first sub-set of stories provides confidence, the second spurs to action.

There is one story common to all the moral theistic religions which has proved of great psychological value in enabling religious men to persevere in carrying out their religious behaviour policies – the story that in so doing they are doing the will of God. And here it may look as if there is an intrinsic connexion between the story and the policy of conduct. But even when the story is literally believed, when it is believed that there is a magnified Lord Shaftesbury who commands or desires the carrying out of the behaviour policy, that in itself is no reason for carrying out the policy: it is necessary also to have the intention of doing what the magnified Lord Shaftesbury commands or desires. But the intention to do what a person commands or desires, irrespective of what this command or desire may be, is no part of a higher religion; it is when the religious man finds that what the magnified Lord Shaftesbury commands or desires accords with his own moral judgement that he decides to obey or to accede to it. But this is no new decision, for his own moral judgement is a decision to carry out a behaviour policy; all that is happening is that he is describing his old decision in a new way. In religious conviction the resolution to follow a way of life is primary; it is not derived from believing, still less from thinking of, any empirical story. The story may psychologically support the resolution, but it does not logically justify it.

In this lecture I have been sparing in my use of the term 'religious belief' (although it occurs in the title), preferring instead to speak of religious assertions and of religious conviction. This was because for me the fundamental problem is that of the meaning of statements used to make religious assertions, and I have accordingly taken my task to be that of explaining the use of such assertions, in accordance with the principle that meaning is to be found by ascertaining use. In disentangling the elements of this use I have discovered nothing which can be called 'belief' in the senses of this word applicable either to an empirical or to a logically necessary proposition. A religious assertion, for me, is the assertion of an intention to carry out a certain behaviour policy, subsumable under a sufficiently general principle to be a moral one, together with the implicit or explicit statement, but not the assertion, of certain stories. Neither the assertion of the intention nor the reference to the stories includes belief in its ordinary senses. But in avoiding the term 'belief' I have had to widen the term 'assertion', since I do not pretend that either the behaviour policy intended or the stories entertained are adequately specified by the sentences used in making isolated religious assertions. So assertion has been extended to include elements not explicitly expressed in the verbal form of the assertion. If we drop the linguistic expression of the assertion altogether the remainder is what may be called religious belief. Like moral belief, it is not a species of ordinary belief, of belief in a proposition. A moral belief is an intention to behave in a certain way: a religious belief is an intention to behave in a certain way (a moral belief) together with the entertainment of certain stories associated with the intention in the mind of the believer. This solution of the problem of religious belief seems to me to do justice both to the empiricist's demand that meaning must be tied to empirical use and to the religious man's claim for his religious beliefs to be taken seriously.

Seriously, it will be retorted, but not objectively. If a man's religion is all a matter of following the way of life he sets before himself and of strengthening his determination to follow it by imagining exemplary fairytales, it is purely subjective: his religion is all in terms of his own private ideals and of his own private imaginations. How can he even try to convert others to his religion if there is nothing objective to convert them to? How can he argue in its defence if there is no religious proposition which he believes, nothing which he takes to be the fundamental truth about the universe? And is it of any public interest what mental techniques he uses to bolster up his will? Discussion about religion must be more than the exchange of autobiographies.

But we are all social animals; we are all members one of another. What is profitable to one man in helping him to persevere in the way of life he has decided upon may well be profitable to another man who is trying to follow a similar way of life; and to pass on information that might prove useful would be approved by almost every morality. The autobiography of one man may well have an influence upon the life of another, if their basic wants are similar.

But suppose that these are dissimilar, and that the two men propose to conduct their lives on quite different fundamental principles. Can there be any reasonable discussion between them? This is the problem that has faced the many moral philosophers recently who have been forced, by their examination of the nature of thinking, into holding non-propositional theories of ethics. All I will here say is that to hold that the adoption of a set of moral principles is a matter of the personal decision to live according to these principles does not imply that beliefs as to what are the practical con sequences of following such principles are not relevant to the decision. An intention, it is true, cannot be logically based upon anything except another intention. But in considering what conduct to intend to practise, it is highly relevant whether or not the consequences of practising that conduct are such as one would intend to secure. As R. M. Hare has well said, an ultimate decision to accept a way of life, 'far from being arbitrary ... would be the most well-founded of decisions, because it would be based upon a consideration of everything upon which it could possibly be founded'.* And in this consideration there is a place for every kind of rational argument.

Whatever may be the case with other religions Christianity has always been a personal religion demanding personal commitment to a personal way of life. In the words of another Oxford philosopher, 'the questions "What shall I do?" and "What moral principles should I adopt?" must be answered by each man for himself'.† Nowell-Smith takes this as part of the meaning of

* R. M. Hare, *The Language of Morals* (1952), p. 69.

† P. H. Nowell-Smith, *Ethics* (1954), p. 320.

morality: whether or not this is so, I am certain that it is of the very essence of the Christian religion.

THE BRAIN AND THE UNITY OF CONSCIOUS EXPERIENCE

Sir John C. Eccles

FRS

THE NINETEENTH

ARTHUR STANLEY EDDINGTON

MEMORIAL LECTURE

15 October 1965

The Brain and the Unity of Conscious Experience

The reality of conscious experience

I BELIEVE THAT THIS problem that I am talking to you about today would have been one of particular interest to Sir Arthur Eddington. Repeatedly in his books he made reference to the problem of consciousness in relation to the physical world that he spoke about with such imagination and such understanding. I can instance his attitude to conscious experience by two brief quotations from his Swarthmore Lecture (1929), *Science and the Unseen World*:

> In comparing the certainty of things spiritual and things temporal, let us not forget this – Mind is the first and most direct thing in our experience; all else is remote inference.
>
> Picture first consciousness as a bundle of sense-impressions and nothing more ... But picture again consciousness, not this time as a bundle of sense-impressions, but as we intimately know it, responsible, aspiring, yearning, doubting, originating in itself such impulses as those which urge the scientist on his quest for truth.

And in his great book *The Philosophy Physical Science* (1939, p. 195) he states:

> The only subject presented to me for study is the content of my consciousness. According to the usual description, this is a heterogeneous collection of sensations, emotions, conceptions, memories, etc. The raw materials of knowledge and the manufactured products of intellectual activity exist side by side in this collection.

And his further statement (p. 206) that 'the unity of consciousness is manifested *because* there are parts for it to unite' is of particular relevance to my theme today.

But Eddington was not alone amongst the great physicists of this century in recognizing the importance and urgency of the problem of consciousness. For example, I can instance Schrödinger's contributions in his monographs, *Science and Humanism* and *Mind and Matter*, and I can give two quotations from a recent lecture, *Two Kinds of Reality*, by Eugene Wigner:

> ...There are two kinds of reality of existence: the existence of my consciousness and the reality or existence of everything else. This latter reality is not absolute but only relative ... Excepting immediate sensations, the content of my consciousness, everything is a construct ... but some constructs are closer, some farther, from the direct sensations. [These constructs are, of course, the physical world.]
>
> As I said, our inability to describe our consciousness adequately, to give a satisfactory picture of it, is the greatest obstacle to our acquiring a rounded picture of the world.

In contrast, until recently it has been fashionable for philosophers and psychologists to discredit or even to deride all problems purporting to derive from the concept of mind or of consciousness. However, the recent reaction to this obscurantism can be illustrated by such books as *The Existence of Mind* (Beloff, 1962) and *On Having a Mind* (Kneale, 1962). Nevertheless, despite this clarification I feel that there is still confusion in the use of such words as mind, mental, mentality, which in some extremely primitive form are even postulated as being a property of inorganic matter! Hence I have refrained from using them, and employ instead either 'conscious experience' or 'consciousness'.

My approach to conscious experience is, in the first instance, based on my direct experience as a conscious self – myself – because I believe this to be the only valid approach to the problem of consciousness. This initial attitude of mine is not solipsistic because similarly I would maintain that each of you has to face up to the problem discussed in this lecture in relation to your own experience of self-consciousness. My conscious experience is all that is given to me in my task of trying to understand myself; and, as I shall argue later, it is only because of and through my experience that I come to know of a world of things and events and so to embark on the attempt to understand it. Furthermore, I have to consider the totality of my conscious experience, not only here and now, but of all my past. Because of the experiences that can be recalled in memory, and so re-experienced, I recognize my unity and identity through all past vicissitudes; it is memory that gives me that continuity of inner experience which belongs to me as a self; and this inner experience comprises not only my memories, but all the sequence of imagery, ideas, desires, volitions and emotional feelings that characterize my waking life, and in addition it includes my dreams and hallucinations.

Sherrington (1940) in his Gifford Lectures (*Man on his Nature*) has written most movingly on the self:

> This 'I', this self, which can so vividly propose to 'do', what attributes as regards 'doing' does it appear to itself to have? It counts itself as a 'cause'. Do we not each think of our 'I' as a 'cause' within our body? 'Within' inasmuch as it is at the core of the spatial world, which our perception seems to look out at from our body. The body seems a zone immediately about that central core. This 'I' belongs more immediately to our awareness than does even the spatial world about us, for it is directly experienced. It *is* the 'self'.

Perceptual experience

In contrast to this inner experience, I have experiences or perceptions that are derived from activation of my sensory receptors. It is solely from such perceptual experiences that I derive the concept of an external world of things and events, which is a world other than the world of my inner experience. Furthermore, it is part of my interpretation of my perceptual experience that my 'self' is associated with a body that is in the objective world; and I find innumerable other bodies that appear to be of like nature. I can exchange communications with them by bodily movements that give rise to perceptual changes in the observer, for example by gestures or, at a more sophisticated level, by speech that is heard or by writing that is read, and thus discover by reciprocal communication that they, too, have conscious experiences resembling mine. Solipsism becomes for me no longer a tenable belief. Eddington (1939, p. 198) makes a valuable statement on this theme:

> Thus recognition of sensations other than our own, though not required until a rather later stage of the discussion, is essential to the derivation of an *external* physical universe. Our direct awareness of certain aural and visual sensations (words heard and read) is postulated to be an indirect knowledge of quite different sensations (described by the words heard and read) occurring elsewhere than in our own consciousness. Solipsism would deny this; and it is by accepting this postulate that physics declares itself anti-solipsistic.

Thus we come to believe that there is a world of selves each with the experience of inhabiting a body that is in a material world comprising innumerable bodies of like nature and a tremendous variety of other living forms and an immensity of apparently non-living matter. I would agree with Wigner that this material or objective world has the status of a second-order or derivative reality. How, it may be asked, can my perceptual experiences give me such an effective knowledge of the objective world that I can find my way round in it and even manipulate it with such success? So effective is this practical operation that I am not conscious of this problem in my whole experience of practical living; my body and its environment appear to be *directly* known to me. This attitude towards perceptual experiences can be termed naive or direct realism, which has of course been rendered untenable by modern neurophysiology.

In response to sensory stimulation, I experience a private perceptual world which must be regarded, neurophysiologically, as an interpretation of specific events in my brain. Hence I am confronted by the problem: how can these diverse cerebral patterns of activity give me valid pictures of the external world? Usually this problem is discussed in relation to visual perception. There seems to be an extraordinary problem in explaining how information from my retinae when relayed to, and activating, my cerebral cortex gives me a picture of the external world with all its various objects in three-dimensional array and endowed with brightness and colour. This epistemological problem has led to much philosophical confusion when it has been discussed on the assumption that fully patterned visual perception is an inborn property of the nervous system. On the contrary, my visual perception is an interpretation of retinal data that in a lifetime of experience I have learned to accomplish, particularly in association both with sensory information provided by receptors in muscles, joints, skin and the vestibular apparatus, and with the central experience of willed effort.

The dependence of perception on active learning

It has of course long been known that there is a remarkable plasticity in the visual perceptive process. I can cite the experiments with inverting prisms. After wearing them for several days and continually practising with active movements, the subject learns to interpret his inverted visual field right way up and can even learn to perform such skilled actions as skiing with such inverted visual data. Many recent experiments of this type by Held and his colleagues have shown that the role of activity is essential in learning to re-interpret distorted visual data.

The most elegant and delightful example of the role of activity in visual learning is provided by the recent experiments of Held and Hein (1963). Litter-mate kittens spend several hours a day in a contraption which allows one kitten fairly complete freedom to explore its environment actively just as a normal kitten. The other is suspended passively in a gondola that by a simple mechanical arrangement is moved in all directions by the exploring litter mate so that the gondola passenger is subjected to the same play of visual imagery as the active kitten, but none of this activity is initiated by the passenger. His visual world is provided for him just as it is for us on a TV screen. When not in this contraption both kittens are kept with their mother in darkness. After some weeks, tests show that the active kitten has learnt to utilize its visual fields for giving it a valid picture of the external world for the purpose of movement just as well as a normal kitten, whereas the gondola passenger has learnt nothing. One simple example of this difference is displayed by placing the kittens on a narrow shelf which they can leave either on one side with a small drop, or on the other side with an intimidating drop. Actually a transparent shelf prevents any untoward damage in getting off on the dangerous side. The actively trained kitten always chooses the easy side, the passive chooses either in random matter.

The conclusion from these and many other experiments on animals and man is that continual-active exploration is essential even if adults are to retain their existing visual discriminations or to learn new ones. The most remarkable physiological and anatomical problems are raised by these intriguing experiments on perception and behaviour, but as yet we can only formulate the problems in the vaguest terms.

These experiments establish that, as a consequence of active or trial-and-error learning, the brain events evoked by sensory information from the retina are interpreted so that they give a valid picture of the external world that is sensed by touch and movement, i.e. the world of visual perception becomes a world in which I can effectively move. It is important to realize that we do not learn from a relaxed kaleidoscope of experiences, but from what we might call 'participation learning'. Actually, this perceptual world is much more synthetic than we imagine; for example, it normally remains fixed and stable when the images on the retina are moved in the most diverse ways by naturally occurring body, head or eye movements, but not, for example, when the eye is moved by an applied pressure. The kinaesthetic information from all these natural movements as well as the sensory information from the vestibular apparatus is synthesized with the retinal information. The action of this automatic correction device for visual perception is best appreciated when there are disturbances of vestibular function; under such conditions there is gross movement of the visually perceived world, which gives rise to the sensation of vertigo.

Thus we can return in a circular manner to the beginning of the story with an understanding of how the external world can be apprehended by means of our sensory mechanisms, receptor organs and their pathways to the brain. Moreover, we can communicate with each other with respect to our interpretation of our sensory experiences and discover that to a very large extent we have agreement with one another on these interpretations which give us what we call the objective world. The measure of this agreement is perhaps best appreciated by reference to situations where there is disagreement. For example, a considerable number of people differ in their interpretation of colour, and we classify them as colour blind or colour defective to varying degrees. Similarly we have 'taste blindness', if we may so call it, of many people to phenylthiocarbamide, which is very bitter to about 75% and tasteless to 25%, and 'smell blindness' of about 18% of males to hydrogen cyanide.

The example of colour blindness illustrates in a most pointed manner the amazing dominance of the majority agreement and the commonsense way of repressing the minority of disagreement. This method works out well for the crude levels of perception using simple criteria of perceptual recognition with nothing more subtle than colour matching; but there are immense divergences in the perceptual experiences of individuals when it comes to such highly sophisticated performances as occur with philosophical arguments, with aesthetic judgements in music, the plastic arts, and literature, and even with such learned skills as tea and wine tasting, and the evaluation of design and décor. And, might I add in a muted tone, these divergences exist amongst scientists in the evaluation and interpretation of experimental data, and in the way in which these data can be used to test scientific hypotheses and so to illuminate our understanding of the natural world. In fact, it is the conflict arising from these differences in interpretation and judgement and belief that gives the drive and zest to our creative adventure and performance in both the sciences and the arts.

Anatomical and physiological basis of conscious experience

As a result of this intensive perceptual training over the years of our lives and of its concentration in the methods of scientific investigations, we have come to learn about sense organs and brains. Gradually by scientific experiments the primitive concepts of their modes of operation in perception have become better understood, both the mode of operation of the sense organs as highly specific detectors of physical or chemical stimuli and the way in which information is communicated as signals (nerve impulses) from them to the cerebral cortex; but you will appreciate that this understanding is performed by highly complex and specific intellectual processes – thinking, observing, assessing, correlating, criticizing, reasoning, imagining.

In this city of Keith Lucas, Adrian, Rushton, Hodgkin, Huxley and Keynes, it should not be necessary for me to tell you anything about the nerve impulse, because it is in Cambridge pre-eminently that the fundamental work has been done on this basic mode of communication in the nervous system. However, I must now give you a brief glimpse into the general structural and functional characteristics of the central nervous system.

The human cerebral cortex is a sheet of about 2000 sq.cm in area and about 3 mm in thickness. It is formed by a very dense packing of nerve cells of many varieties and sizes, in all about ten thousand million, and it is often likened to a vast telephone exchange. Within the last five years there have been enormous advances in studying the cerebral cortex by electron microscopy and in employing both intracellular and extracellular recording from the pyramidal cells in order to study the way in which the nerve cells communicate with each other by means of the synaptic contacts. An impulse discharged from one cell causes a momentary activation of the many excitatory or inhibitory synapses that each cell forms with other cells, often many hundreds. Some speculative glimpse of neuronal operation can be achieved by realizing that many almost synchronous excitatory synaptic bombardments are essential for causing any cell to generate an impulse and itself thus to contribute to the further spread of neuronal activity. For an effective spread of activity each neurone must receive synaptic activation probably from hundreds of neurones and itself transmit to hundreds of others. One is thus introduced to the concept of a wave-front comprising a kind of multi-lane traffic in hundreds of neuronal channels, so that the wave-front would sweep over at least 100,000 neurones in one second. Furthermore, there is a great deal of evidence to show that a particular neurone may participate in the patterns of activity developing from many different inputs.

However, as a result of this immense scientific study, we are still only at the first stage of the events concerned in perceptual awareness and have virtually not approached at all the more complex problems of perceptual recognition and judgement. There is much neurophysiological evidence that a conscious experience arises only when there is some specific cerebral activity. For every experience it is believed that there is a specific spatio-temporal pattern of neuronal activity in the brain (cf. Fessard, 1961). Thus with perception the sequence of events is that some stimulus to a sense organ causes the discharge of impulses along afferent nerve-fibres to the brain, which, after various synaptic relays, eventually evoke specific spatio-temporal patterns of impulses in the neuronal network of the cerebral cortex. The transmission from sense organ to cerebral cortex is by a coded pattern of nerve impulses that is quite unlike the original stimulus to that organ, and the spatio-temporal pattern of neuronal activity that is evoked in the cerebral cortex would be again different. Yet, as a consequence of these cerebral patterns of activity, I experience sensations (more properly the complex constructs called percepts) which in my private perceptual world are 'projected' to somewhere outside the cortex; it may be to the surface of the body or even within it, or, as with sight, hearing or smell, to the outside world. However, as succinctly expressed by Russell Brain (1951), 'the only necessary condition of the observer's seeing colours, hearing sounds, and experiencing his own body is that the appropriate physiological events shall occur in the appropriate areas of his brain.'

It cannot be too strongly emphasized that this investigation into the neuronal mechanisms of the cerebral cortex is still at an extremely primitive stage and hence gives but some dim and shadowy picture of the amazing intricacy of pattern woven in space and time by the sequential activation of neurones in multi-lane traffic over the ten thousand million components in the cortical slab of cells. It has been surmised that many millions of cells take part in the simplest cortical response.

We can further speculate that the human cerebral cortex surpasses that of all other animals in its potentiality for developing subtle and complex neuronal patterns of the utmost variety, for from this would stem the richness of human performance as compared with that of even the most intelligent animal.

The threshold of conscious experience

It has long been known that sensations can be evoked by electrical stimulation of the brain of conscious subjects, and a most thorough investigation has been made by Penfield and his associates (Penfield and Jasper, 1954). Usually these sensations are disordered experiences or paraesthesia; light or colours from the visual area, tingling and numbness from the somaesthetic area (the area concerned in general body sensations); noises from the auditory area. Recently Libet and his colleagues (1965) have utilized these responses of the somaesthetic area in an attempt to discover the nature of the neuronal activity that leads to a conscious experience. They applied very weak trains of brief electric pulses usually at frequencies of 30–60 a second to the exposed cerebral cortex of conscious subjects who had generously volunteered time during some therapeutic brain surgery. The object of the experiment was to determine the stimulus that just sufficed to cause them to report that they had a conscious experience, which was of course of a somaesthetic character, usually abnormal, but in about one third it was normal – for example, a sensation of pressure or touch or movement, and even heat or cold. It was of great interest that, as the number of stimuli in the train increased, there was a large reduction in the strength required to produce a conscious experience, and that at least half a second of repetitive stimulation was required for the weakest stimulus. Continuation of the weakest stimulus beyond the time for production of a sensation did not increase the sensation, but merely prolonged it at the threshold of feeling.

There would be general agreement that each electrical stimulus of the train would be exciting the discharge of impulses from nerve cells and that the effect of duration in lowering threshold strength indicates that there must be an elaboration of the spatio-temporal patterns of impulse discharges before a conscious experience arises. Furthermore, it is suggested that, with all conditions of threshold stimulation, there is a delay of at least half a second before the onset of the experienced sensation. Evidently there is opportunity for a great elaboration of neuronal activity in complex spatio-temporal patterns during the 'incubation period' of a conscious experience at threshold level.

These same temporal characteristics are exhibited with stimulation applied to subcortical white matter or to the thalamus, so it can be assumed that there is an essential factor of elaboration of activated neuronal patterns before a sensation is experienced. This inference conforms with the finding of Jasper (1965) that the initial electrical responses produced by afferent volleys to the cortex are not related to conscious experience. For example, these initial responses are unaltered in relatively deep anaesthesia, but following the initial response there are small after-waves for a second or more and these are very sensitive to the depth of anaesthesia, and are in fact correlatable with the experiencing of sensation.

With the visual system also there is evidence that at least 0.2 second of cortical activity is required before a just-threshold flash of light can be detected. This elaboration time for a conscious experience may be as long as 1 second; so a sensory input may evoke quick motor reactions by operation at subconscious levels before it is actually experienced. It is important to recognize that measurements of reaction time cannot be used as a measure of the time required for elaboration of a conscious experience.

The neuronal activity concerned in conscious experience

The time of at least 1/5th of a second for the elaboration of the neuronal substrate of a conscious experience is very long indeed. The time for transmission from one nerve cell to another is no longer than 1/1000th of a second; hence there could be a serial relay of as many as 200 synaptic linkages between nerve cells before a conscious experience is aroused. Many thousands of nerve cells would be initially activated, and each nerve cell by synaptic relay would in turn activate many nerve cells. The immensity of this patterned spread throughout the neuronal pathways of the brain is beyond all imagining. This tremendous complication of neuronal activity in my brain is required before a sensory input is perceived by me even in the rawest form; and responses involving comparison, value, judgement, correlations with remembered experiences, aesthetic evaluations undoubtedly take much longer, with the consequence that there must be quite fantastic complexities of neuronal operation in the spatio-temporal patterns woven in the 'magic loom', to use the phrase of Sherrington. For example, Mountcastle (1965) states that:

> The tide of experiment moves towards an elucidation of the time-dependent, dynamic aspects of cortical function. One may suppose, and indeed some observations already indicate, that the sequential forming and reforming of new and highly complex patterns of activity, occupying both the cortical cells activated initially by a sensory stimulus and others independent of it, results in functional patterns far more complex than those predicted by the columnar organization alone, and it is this aspect of neural activity to which I believe we must look for the neural correlates of the perceptive process.

Moruzzi (1965) presents an impressive amount of data in support of the postulate that only an extremely small fraction of all the sensory input is actually experienced – for example, during sleep there is a continuing overall activity of cortical nerve cells, and Evarts (1964) finds that there may even be increased activity of some cortical neurones. Again it will be recognized that only a very small fraction of the immense volume of visual information is utilized in visual perception from moment to moment. Moruzzi concludes that only an extremely small proportion of all the patterned neuronal activity that is going on in a brain at any one time gives rise to conscious experience, though, within limits, we may apparently direct our attention to other neuronal patterns which, as a consequence, are then consciously experienced. Likewise, only an extremely small fraction of the sensory input into the brain can be recalled in the process of memory even for the few minutes of the short-term memory. Moruzzi (1965) concludes that

> all these considerations lead to the conclusion that the neural processes underlying learning and forgetting, storage and retrieval of memory traces are quantitatively small with respect to the background activity

of the cerebrum, although the highest achievements of mankind, from artistic creation to scientific discovery, are dependent upon them.

The great sub-cortical nuclei, and particularly the reticular activating system, have been shown by Magoun and Moruzzi and by Bremer to arouse or energize the cortex by a continual barrage of impulse discharge, so maintaining levels of cortical awareness. However, though this background or supporting activation is essential to the maintenance of consciousness, it must be distinguished from the process of attention which operates in some more selective manner than can be provided by these non-specific systems. In agreement with Mountcastle, I would regard all the fine grain of conscious perception that we experience during attention as essentially arising in the cortex.

Unity of conscious experience and the cerebral commissures

There are remarkable problems raised by the fact that we have two cerebral hemispheres, each with an immense amount of localized performance, with inputs of general body sensation and vision channelled into the one or other side, and with movement likewise dependent on the one or the other side of the motor cortex; yet we experience what Bremer (1965) aptly calls a 'mental unity'. We can ask: how can the diversity and the tremendous dispersion of activity in the spatio-temporal patterns of the brain give rise to this unity and, from moment to moment, the relative simplicity of our conscious experience so that the play of experience appears to be, as it were, all on the stage before one single conscious self?

Undoubtedly the neurological correlate of this unification of experience arising from neural events in the two cerebral hemispheres is to a large extent the enormous commissural tract, the corpus callosum that links the mirror-image areas of the two hemispheres; and to a lesser extent there is also the commissural linkage by the anterior commissure and the massa intermedia. As is well known, this mental unity in man remains intact after large lesions or surgical destructions of the cerebral hemispheres, even the integration areas for symbolic expression in language being destroyed; and we can all experience in dreams the fragmentary and chaotic imagery which is part of our experiencing unitary self. Perhaps even more remarkable are the observations of Penfield, who was able to evoke remote audio-visual memory of illusory character by electrical stimulation of the temporal lobe; yet these strange experiences were assimilated by the subject and recognized as remembrances of long-forgotten incidents in the lifetime of the same self. The same extraordinary ability of the experiencing self to build a unity out of diversity is illustrated by the phenomenon of drug-induced hallucination. No matter how bizarre the experiences, they are recognized as belonging to the self and not due to some privileged view of mental happenings in some other self – or some amputated component of the original self, i.e. there is no mental diplopia.

This postulate of the key role of the brain commissures in linking the cerebral hemispheres has been tested in the last decade by Myers and Sperry and their associates (Myers, 1961; Sperry, 1964, 1965) in experiments of remarkable ingenuity. In the cat and monkey they have split the optic chiasma, so that each eye feeds into a cerebral hemisphere on its own side. In the split-chiasma animals what was learned by visual inputs into one hemisphere was transferred to the other hemisphere and laid down as a memory trace, where it could be detected after a subsequent splitting of the brain; that is, the information that goes in from one eye to one hemisphere is communicated to the other hemisphere at the time of the laying down of memory traces. By contrast, after the brain was split, there was no transfer of learning from one to the other side. The two sides of the brain could be trained to give diametrically opposite responses to stimuli. At this level of testing, the animal is divided into two independently learning and behavioural organisms. By contrast, if this attempt to build up opposed responses to information fed into the two eyes of the split-chiasma animal was made before the splitting of the brain, a severe behavioural conflict was aroused in the animal. This behavioural conflict was also observed when attempts were made to train animals in an opposed manner with signals involving touch and kinaesthesis. It can be concluded that the brain commissures are essentially concerned in the transfer of information between the two hemispheres, so that they can share in learning and memory. The neurological substrate of learning that is called an engram (Lashley, 1950), is normally laid down in both hemispheres of the cat. Experiments on monkeys and anthropoid apes reveal that this duality of memory trace is there a less prominent feature. And with man there are remarkable examples of complex memories restricted to one hemisphere, as, for example, occurs with language in the dominant hemisphere, which is the left in a right-handed subject.

The most remarkable examples of experiment on interhemispheric communication have been made on two human subjects in which a surgical separation of the two cerebral hemispheres was made in order to control intractable epilepsy, and mercifully this operation of severing the corpus callosum, the anterior commissure and the massa intermedia was successful therapeutically (Sperry, 1964, 1965). Just as with the split-brain animals, these subjects display no gross signs of incoordination of response, nor do they experience any splitting of their mental unity, such as might be called a mental diplopia. However, they reveal gross disorders both of reaction and of experience when tested appropriately. The most remarkable findings stem from the almost invariable unilateral representation of language in the dominant cortical hemisphere, which is the left in both these cases. For example, they are unable to read with the left half of the visual field, which feeds exclusively into the right hemisphere (the minor hemisphere), and commands conveyed verbally are carried out with the right side only. They react to stimuli applied to the left visual field, sometimes appropriately, but without being able to give an account of what they are doing. Similarly, they have no detailed knowledge of touch or movements on the left side, and if blindfolded they do not know what the left side is doing. Evidently, the dominant hemisphere of the brain neither 'knows' nor 'remembers' the experiences and activities of the other hemisphere. The difficulties of these subjects derive from the uncontrollable behaviour of the left side, particularly the left hand. Often they try to control the left hand by their right hand.

All the evidence produced by these two cases is explicable by the postulate that, when bereft of commissural linkages with the dominant hemisphere, the minor hemisphere behaves as a computer with inbuilt skills of movement, with recognition of the form and function of objects, and with the ability to learn; nevertheless, the dominant hemisphere with its ability of linguistic expression remains oblivious of all this performance. For example, as stated by Sperry (1965),

> ...the subject may be blindfolded and some familiar object like a pencil, a cigarette, a comb or a coin placed in the left hand. Under these conditions the mute hemisphere connected to the left hand feeling the object perceives and appears to know quite well what the object is. Though it cannot express this knowledge in speech or in writing, it can manipulate the object correctly, it can demonstrate how the object is supposed to be used, and it can remember the object and go out and retrieve it with the same hand from among an array of other objects either by touch or by sight. While all this is going on, the other hemisphere meanwhile has no conception of what the object is, and, if asked, says so. If pressed for an answer, the speech hemisphere can only resort to pure guess-work. This remains the case just so long as the blindfold is kept in place and/or other avenues of sensory input from the object to the talking hemisphere are blocked. But let the right hand cross over and touch the test object in the left hand; or let the object itself touch the face or head as in demonstrating the use of a comb, a cigarette or glasses; or let the object make some give-away sound, like the jingle of a key case, then immediately the speech hemisphere also comes across with the correct a answer.

Likewise, happenings in the left visual field remain unknown to the dominant hemisphere. For example, Sperry (1965) reports:

> The subjects fail on such simple tasks, for example, as that involved in discriminating whether red and green half fields presented together are the same or different in colour where the response involves only a simple nodding or shaking of the head 'yes' or 'no' – in other words, with everything favouring any cross integration that might be present. The same task caused no difficulty to either hemisphere when the two colours or other stimuli were presented within the same half retinal field and hence projected to the same hemisphere. Comparison of the directional tilt of broad straight lines running across the visual field and interrupted in the centre of the screen went easily again when both parts of the bar fell in one field, but when the two parts fell in right and left fields separately, the subjects were unable to indicate whether the two bars were lined up straight across the midline or at an angle. When the response in this test involved manual copying of the perceived lines, the initial result was for each hand to record and draw only the part of the line within its own half of the visual field, the other half being omitted. When both hands had a pencil and worked simultaneously, both parts of the line were drawn correctly, indicating double simultaneous perception and response.

We can summarize this by stating that the goings-on in the minor hemisphere, which we may refer to as the computer, never come into the conscious experience of the subject.

It is particularly interesting to consider the problem of freewill in relation to these two separated cerebral hemispheres. So far as the dominant hemisphere and the right side of the body are concerned the situation is exactly as for a normal patient. However, the conscious subject has no direct control over what the left hand is doing, nor is there any cognizance of these movements except in so far as they produce information that is channelled by neural pathways into the dominant hemisphere, for example from the right visual field or with palpation by the right hand. We have already seen that the left hand can perform appropriate actions of recognition and that these are quite unknown to the conscious patient; in fact the dominant hemisphere may try, by using the right hand, to interfere with the correct responses being made by means of the minor hemisphere in response to information fed into it by the left visual field.

With progressive recovery from the operation, one of the subjects was able to develop some voluntary control of the left hand, but this was shown to be due entirely to the uncrossed pyramidal pathway from the left or dominant hemisphere directly to the nerve cells controlling the muscles of the left arm, and not at all to a conscious voluntary action exerted by the minor hemisphere, or computer. Yet there is plenty of evidence that this hemisphere does have a complex reacting life all of its own, though it has no power of overt expression because linguistic expression lies entirely with the dominant hemisphere. For example, by means of the minor hemisphere the left hand can point out a matching picture from many others that have been flashed into the left visual field. It can pick up the correct written name of an object flashed on the screen, or, *vice versa*, it can read a name and retrieve the designated object. For example, the words cup, fork or apple flashed on the screen cause the left hand to pick up the appropriate object that is also in the left visual field. All this occurs with the subject having no conscious experience whatever of what is happening with all these performances of the minor hemisphere, which in this way is displaying perception, comprehension, reading, retrieving and learning.

Sperry (1965) argues that the presence of consciousness in a hemisphere would not be demonstrable in the absence of some appropriate linguistic mode of expression and therefore that the minor hemisphere may in fact be responsible for conscious states which cannot be indicated to the observer because of the failure of symbolic communication in language. In fact we can agree that the problem of trying to discover if the activities of the minor hemisphere are actually resulting in conscious experiences is equivalent to our problem of trying to discover if an animal's cerebral activity gives rise to conscious experiences. Sperry even suggests that at some later date by more devious pathways, as for example the long-loop reticular pathways suggested by Bremer, it may be possible for the minor hemisphere to communicate with the dominant hemisphere and so be able to achieve linguistic expression. In that case it might even be able to report remembered experiences from the present experiments, though in the absence of present report it might be concluded that there were no such experiences!

An analogous situation can occur in the so-called automatic states of a subject with normal interhemispheric communication. Last century the great English neurologist, Hughlings Jackson, made a special report of mental disorders which are now recognized as arising from localized epileptic seizures in the amygdaloid area. In these automatic states, persons can, for quite long periods, react in a complex and sophisticated manner and yet, on recovery, have no trace of any recollection. For example, a doctor in an automatic state examined a patient and made reasonably accurate notes of the examination, yet remembered nothing afterwards. The question may be asked: was he conscious during this automatic state? My answer is that we cannot assume this to be so if the subject has no trace of any remembered experiences. In the absence of any evidence we must be agnostic, just as with the question of consciousness in animals.

Thus it is clear that the loss of commissural communication between the two halves of the brain has caused a split both in the perceptual and in the operational functioning of the person. The really remarkable finding is that the conscious self, with all its linguistic and sophisticated behavioural performance, seems to be represented solely in the dominant hemisphere in these split-brain patients. Unity of conscious experience is retained at the expense of a loss of all the experience that would be expected to be associated with the activities of the minor hemisphere. One may well wonder if this is the case in normal subjects, so that information fed into this hemisphere reaches consciousness only after interhemispheric transfer. This suggestion gives rise to the further question of what is the functional importance of the minor hemisphere other than to receive from the sense organs and transmit to the dominant hemisphere, and in turn to receive from this hemisphere and transmit to the muscles of the opposite side. It should be mentioned, however, that there is evidence that motor skills and construction of spatial relations and perspective in drawing were better performed by the left hand and the minor hemisphere.

In conclusion I feel a profound dissatisfaction when I contemplate the present attempts to account for the undoubted unity of my conscious experience. One is confronted by the extraordinary problem of trying to reconcile the unitary nature of my conscious self with the neurological events of the utmost diversity and complexity that are assumed to underlie it, and that involve the 'weaving' by impulses of spatio-temporal patterns in the 'magic loom' of Sherrington with its thousands of millions of units or nerve cells. Penfield and Jasper (1954) attempt to soften this antithesis by postulating that the unification of experience occurs in the centrencephalic system which includes the great subcortical nuclei that give to and receive from all parts of the wide-spreading cerebral cortex. To them there is a relative simplicity in conscious experience in contrast to the enormous complexity of the sensory input. Yet I would argue that this simplification occurs only in respect of the moment-to-moment perceptions that we experience. By attention and concentration we can greatly sharpen the focus of experience and perceive a fineness of grain matching the information that the sense organs feed into the brain. In any case the problem of the unity of experience still remains an enigma whether the neural substratum is spatio-temporal patterns of neuronal activity in these large subcortical nuclei of the centrencephalic system, or in the cortex itself. As Sherrington (1940) points out, there is no 'centralization upon one pontifical nerve cell'. The antithesis must remain that our brain is a democracy of ten thousand million nerve cells, yet it provides us with a unified experience.

Does the uniqueness of the experiencing self derive from genetic uniqueness?

We may take it as certain that my conscious self depends uniquely on my brain and not on other brains. I think telepathy is still a tenable belief; but, if it exists at all, it provides an extremely imperfect and inefficient way of transferring information from the neural activity of one brain and its associated conscious experiences to the conscious experiences arising from my brain. This unique interdependence between a brain and a conscious self raises a problem that has always been of great interest to me. It has been expressed by the great American biologist, H. S. Jennings (1930), in a speculative chapter entitled 'Biology and Selves' in his book *The Biological Basis of Human Nature*. However, the climate of opinion has been so unfavourable to such speculations that Jennings's ideas have been almost universally neglected. Yet they are very relevant to the problems raised in this lecture: namely, the uniqueness of the conscious experiences that each of us enjoys, and their relationship to the neuronal activities of our brains.

Two questions may be asked: what is the nature of this consciously experiencing self? and how does it come to be related in this unique manner with a particular brain? I am aware that to many these are not valid questions. My only rejoinder can be that to me they are the most fundamental and important questions that can be asked; and let me state, as a brief autobiographical aside, that I have held this belief since I was 18 years old, when I had a kind of sudden illumination of these problems, and I have been driven on by their interest and urgency to spend my life studying the nervous system.

Jennings formulated with a masterly and lucid style two problems that to him were quite unanswerable. Both were related to the superficially attractive hypothesis that the uniqueness of the self derives from the uniqueness of the particular gene combination belonging to that self, or, as Jennings expresses it, 'the assumption that it is diversity of gene combination that gives origin to distinctiveness of selves'.

In the first place, of course, that assumption is refuted by the distinctiveness that is experienced by identical twins with their identical gene combinations. Alike as these twins are to external observers, each in its own conscious experiences and self-hood is as distinct from its fellow twin as it is from any other self. Evidently, identity of gene combinations must be compatible with distinctiveness of experiencing selves.

The second problem has a universal reference to all conscious selves, to each one of us. It was formulated by Jennings in relation to the genetic theory that any individual (except identical twins) genetically is a unique and never-to-be-repeated knot of strands of genes that has come by inheritance through countless individuals from the remote past. Jennings asks:

> What is the relation of my self, identified as it is with one particular knot in the great network that constitutes humanity, to the other knots now existing? Why should I be identified with one only? To an observer standing apart from the net, it will not appear surprising that the different knots, since they are formed of diverse combinations of strands, should have different peculiarities, different characteristics. But that the observer himself – his total possibility of experience, that without which the universe for him would be non-existent – that he himself should be tied in relations of identity to a single one of the millions of knots in the net of strands that have come down from the unbeginning past – this to the observer appears astonishing, perplexing. Through the operation of what determining causes is my self, my entire possibility of experiencing the universe, bound to this particular one of the combination of strands, to the exclusion of some millions of others? Would *I* never have been, would *I* have lost my chance to participate in experience, would the universe never have existed for me, if this particular combination had not been made?

If the existence of *me* is thus tied to the formation of a particular combination of genes, one may enter upon calculations as to the chances that I should ever have existed. What are the chances that the combination which produced me should ever have been made? If we depend on the occurrence of the exact combination of genes that as a matter of fact produced us, the odds are practically infinite against your existence or my existence.

And what about the selves that would have come into existence if other combinations of genes had been made? If each diverse combination yields a different *self*, then there existed in the two parents the potentialities, the actual beginnings, of thousands of billions of selves, of personalities, as distinct as you and I. Each of these existed in a form as real as your existence and my existence before our component germ cells have united. Of these thousands of billions, but four or five come to fruition. What has become of the others?

And of course to go further backwards in our genetical tree makes the problem even more preposterously fantastic. Hence on both these grounds, I must reject this materialistic doctrine that the uniqueness of my conscious experiencing self is derived from the uniqueness of my genetic make-up. What then determines the uniqueness of my self?

I have found that a frequent and superficially plausible answer to this question is the statement that the determining factor is the uniqueness of the accumulated experiences of a self throughout its lifetime. And this factor is also invoked to account for the distinctiveness of uniovular twins despite their genetic identity. It is readily agreed that my behaviour, my character, my memories, and in fact the whole content of my inner conscious life are dependent on the accumulated experiences of my life; but no matter how extreme the change that can be produced by the exigencies of circumstance, I would still be the same self able to trace back my continuity in memory to my earliest remembrances at the age of one year or so, the same self in a quite other guise. Thus the accumulated experiences of a lifetime cannot be invoked as the determining or generating factor of the unique self, though naturally they will enormously modify all the qualities and features of that self. The situation is analogous to the Aristotelian classification into substance and accidents.

Jennings must have appreciated the fallacy of attempting to derive the uniqueness of self from the experiential history of an individual, for in searching for an explanation he develops the following remarkable speculations:

> To work this out in detail, one would apparently have to hold that the human self is an entity existing independently of genes and gene combinations; and that it merely enters at times into relations with one of the knots formed by the living web. If one particular combination or knot should not occur, it would enter into another. Then each of us might have existed with quite different characteristics from those we have – as our characteristics would indeed be different if we had lived under different environments ... It could be held that there is a limited store of selves ready to play their part, that the mere occurrence of two particular cells which may or may not unite has no determining value for the existence of these selves, but merely furnishes a substratum to which for reasons unknown they may become temporarily attached... And what interesting corollaries might be drawn from such a doctrine, as to the further independent existence of the selves after the dispersal of the gene combinations to which they had been attached! Certainly no one can claim that biological science establishes or indeed favours that doctrine. But since biology itself furnishes no positive doctrine of the relation of selves to gene combinations, the question is a fair one: Does biological science make the holding of that doctrine impossible?

General conclusions

It is important to recognize that in the first instance this question of the relation of a self to gene combinations can be asked only by an experiencing self of its own existence. For example, I can ask it in relation to my own self, and I reply that I must face up to the problems of my own personal existence as an experiencing self that is dependent on the functioning of a brain, which I try to understand as a biological mechanism; and that my brain has had a biological origin as a consequence of a gene combination and the ensuing embryological development. My experiencing self is the only reality I know by direct apprehension – all else is a second-order or derivative reality. The arguments presented by Jennings preclude me from believing that my experiencing self has an existence that merely is derivative from my brain with its biological origin, and with its development under instructions derived from my genetic inheritance. Nor do I believe with the physicalists that my conscious experiences are *nothing but* the operation of the physiological mechanisms of my brain. It may be noted in passing that this extraordinary belief cannot be accommodated to the fact that only a minute amount of cortical activity finds expression in conscious experience. Contrary to this physicalist creed, I believe that the prime reality of my experiencing self cannot with propriety be *identified* with some aspects of its experiences and its imaginings – such as brains and neurones and nerve impulses and even complex spatio-temporal patterns of impulses. The evidence presented in this lecture shows that these events in the material world are necessary but not sufficient causes for conscious experiences and for my consciously experiencing self.

If we follow Jennings, as I do, in his arguments and inferences, we come to the religious concept of the soul and its special creation by God. I believe that there is a fundamental mystery in my existence, transcending any biological account of the development of my body (including my brain) with its genetic inheritance and its evolutionary origin; and, that being so, I must believe similarly for each one of you and for every human being. And just as I cannot give a scientific account for my origin – I woke up in life, as it were, to find myself existing as an embodied self with this body and brain – so I cannot believe that this wonderful divine gift of a conscious existence has no further future, no possibility of another existence under some other unimaginable conditions. At least I would maintain that this possibility of a future existence cannot be denied on scientific grounds.

For a final statement of my belief, I would like to quote from an earlier Eddington lecture by Thorpe (1961):

> I see science as a supremely religious activity but clearly incomplete in itself. I see also the absolute necessity for belief in a spiritual world which is interpenetrating with and yet transcending what we see as the material word ... Similarly I believe that anyone who denies the validity of the scientific approach within its sphere is denying the great revelation of God to this day and age. To my mind, then, any rational system of belief involves the conviction that the creative and sustaining spirit of God may be everywhere present and active; indeed I believe that all aspects of the universe, all kinds of experience, may be sacramental in the true meaning of the term.

REFERENCES

A. S. Eddington, *Science and the Unseen World*. London: George Allen and Unwin Ltd (1929).

A. S. Eddington, *The Philosophy of Physical Science*. London: Cambridge University Press (1939).

E. Schrödinger, *Science and Humanism*. London: Cambridge University Press (1951).

E. Schrödinger, *Mind and Matter*. London: Cambridge University Press (1958).

J. Beloff, *The Existence of Mind*. London: Macgibbon and Kee (1962).

W. Kneale, *On Having a Mind*. London: Cambridge University Press (1962).

C. Sherrington, *Man on his Nature*. London: Cambridge University Press (1940).

R. Held and A. Hein, 'Movement-produced stimulation in the development of visually guided behaviour', *J. comp. physiol. Psychol.* no. 56, 872-6 (1963).

A. Fessard, 'The role of neuronal networks in sensory communications within the brain'. In *Sensory Communication*. Ed. W. A. Rosenblith. London: John Wiley and Sons Ltd. London (1961).

W. Russell Brain, *Mind, Perception and Science*. Oxford: Blackwell Scientific Publications (1951).

W. Penfield and H. Jasper, *Epilepsy and the Functional Anatomy of the Human Brain*. Boston: Little, Brown and Co. (1954).

B. Libet, 'Brain stimulation and the threshold of conscious experience. In *Brain and Conscious Experience*. Ed. J. C. Eccles. Heidelberg: Springer-Verlag (1965).

H. H. Jasper, 'Pathophysiological studies of brain mechanisms in different states of consciousness'. In *Brain and Conscious Experience*. Ed. J. C. Eccles. Heidelberg: Springer-Verlag (1965).

V. B. Mountcastle, 'The neural replication of sensory events in the somatic afferent system'. In *Brain and Conscious Experience*. Ed. J. C. Eccles. Heidelberg: Springer-Verlag (1965).

G. Moruzzi, 'The functional significance of sleep with particular regard to the mechanisms underlying consciousness'. In *Brain and Conscious Experience*. Ed. J. C. Eccles. Heidelberg: Springer-Verlag (1965).

E. V. Evarts, 'Temporal patterns of discharge of pyramidal tract neurons during sleep and waking in the monkey'. *J. Neurophysiol.* no. 27, 152-71 (1964).

F. Bremer, 'Neurophysiological correlates of mental unity'. In *Brain and Conscious Experience*. Ed. J. C. Eccles. Heidelberg: Springer-Verlag (1965).

R. E. Myers, 'Corpus callosum and visual gnosis'. In *Brain Mechanisms and Learning*. Ed. J. F. Delafresnaye. Oxford : Blackwell Scientific Publications (1961).

R. W. Sperry, 'The great cerebral commissure'. *Scientific American*, no. 210, 42-52 (1964).

R. W. Sperry, 'Hemispheric interaction and the mind-brain problem'. In *Brain and Conscious Experience*. Ed. J. C. Eccles. Heidelberg: Springer-Verlag (1965).

K. S. Lashley, 'In search of the engram'. *Symp. Soc. exp. Biol.* London: Cambridge University Press (1950).

H. S. Jennings, *The Biological Basis of Human Nature*. New York: W. W. Norton and Co., Inc. (1930).

W. H. Thorpe, *Biology, Psychology and Belief*. London: Cambridge University Press (1961).

I BELIEVE...

Dame Kathleen Lonsdale

DBE DSc FRS

THE EIGHTEENTH

ARTHUR STANLEY EDDINGTON

MEMORIAL LECTURE

6 November 1964

I Believe …

I NEVER KNEW ARTHUR Stanley Eddington as a person. This was in a way surprising. I took my first degree in Physics, in the University of London, when Eddington was 40 years old. The General Theory of Relativity had been formulated only five years earlier and Eddington was probably its greatest interpreter to the English-speaking world. Yet I have no recollection of ever having heard him lecture, let alone of having met him. I might well have met him in another capacity, for I joined the Society of Friends in 1935, nine years before he died. But during those nine years I was engaged in the treble job of running a home, bringing up three children and carrying on full-time research work at the Royal Institution as a physicist and crystallographer. This involved a degree of organisation that included responsibility and service in my local Friends' Meeting but which left no time for the centralised Friends' activities in which I might perhaps have met Eddington. I very much regret not having done so; although I am told that he was so shy that he was not an easy man to meet or to talk with. Yet he did not write as if he were shy: he had the gift of a direct, almost conversational approach that made difficult ideas seem simple. He helped me very greatly.

Now the autobiographical detail with which I have begun my lecture is not irrelevant to the subject I have chosen: 'I believe … '. These rather personal words are fundamental both to scientific research and to religious seeking. I shall ask you to consider whether we use them in a different way in the two cases. They are even more fundamental to the bringing up of small children. And here I must certainly explain myself if I am not to be seriously misunderstood. Perhaps I can best do so by going even further back into my autobiographical story, because what we come to believe as adults is certainly conditioned in a positive or negative way by our background. My mother was a Baptist; not a Strict Baptist in the sectarian sense, but certainly a convinced fundamentalist. My father was an agnostic, and he had no patience with my mother's religion; in particular, he strongly objected to her giving away any money (of which we had little enough) for religious purposes. I loved and respected my mother. I was a little afraid of my father's temper and I had very little affection for him, although looking back I can see that he was hardworking and upright in many ways. His influence in my life was a curious one. I became a lifelong teetotaller largely because he was not. I have never even attempted smoking principally because he smoked so much. But he was a great reader and a natural mathematician. Our house was full of encyclopaedias to dip into and of books worth reading; and my scientific turn of mind I believe I owe to him.

In the little town in Southern Ireland where I was born and where I lived for the first five years of my life there was no Baptist Church; so mother took us to the Church of Ireland services and we attended a Methodist Sunday School. I can fairly claim to have had an ecumenical upbringing! But what was impressed upon me well before I was 10 years old was that unless I accepted certain statements of religious doctrine as literal fact then I was not a Christian, and that unless I was 'saved' I was damned. The pressure was too much for a sensitive child. Before I went to a Secondary School I made a 'confession of faith' and I was subsequently baptised into membership of the Baptist Church. I have the Bible, now falling to pieces, that was given to me on that occasion more than half a century ago. But all the time, at the back of my mind and to my great distress, a small voice was saying 'It's not true, you know. A really loving Heavenly Father would not condemn hundreds of thousands of people who simply *can't* believe in something that just isn't so.' Throughout adolescence I continued to question and to doubt and yet to feel rather unhappy because to doubt implied disloyalty to my mother, of whom I was so fond. The doubts went deeper, too. I had jibbed first at *having* to believe anything, at being obliged to accept the truth of any statement upon authority only. Perhaps my scepticism was precocious; it was certainly involuntary. I would have preferred to be credulous; it would have been far less strain.

I was growing up in an exciting time. The Wright brothers had designed and flown the first heavier-than-air machine in the year in which I was born. Road transport was becoming revolutionised by mechanisation. Incandescent gas mantles and then electric light bulbs were slowly displacing candles and paraffin lamps and batswing burners. Magic lanterns and the silent cinematograph were being played to full houses. My elder brother worked in one of the first wireless stations established in the south-west of Ireland to pick up signals from ships on the Atlantic. He gave me a home-made cat's-whisker receiving set with ear-phones. The special theory of relativity and the quantum theory were formulated in the first decade of the twentieth century and the diffraction of X-rays by crystals, discovered in 1912, promised to give new experimental insights into the structure of matter on the one hand and the nature and properties of electromagnetic radiation on the other. Knowledge was expanding at a breathtaking rate, in spite of the interruption of the First World War. Many of the technical discoveries being made had become possible only because of the new scientific attitude which did not hesitate to challenge traditional thought and to test theory by planned experiment. I could believe in the possibility that one day I might myself fly, because I could see manned flying-machines overhead, although every one was a marvel in those days and there were plenty of people who said that if God had intended us to fly he would have given us wings. But although flying seemed and was a miracle of faith, ingenuity and persistence, it was a very different sort of miracle from the Virgin Birth, the Incarnation or the Resurrection. Not only did these once-for-all events strain my credulity and loyalty to breaking-point, but I began to wonder whether it was even honest to remain loyal to the idea of a God whom I could not define, and whom no one could satisfactorily define for me.

What really prevented me from becoming an acknowledged agnostic, however, let alone an atheist, was the difficulty I experienced in finding *any* philosophy of life that could satisfy my longing to know the truth, and the fact that I preferred those of my acquaintances and those of whom I had read who were professed Christians, to those who were antagonistic to religion. In particular, I greatly respected the teachings of Jesus and admired his life. I was beginning to question, also, whether in fact Jesus had made or had meant to make those claims for himself, that his followers had made for him. I joined Sir William Bragg's research group in 1922 and as a research scientist I soon began to realise that there were many scientific concepts that could not be satisfactorily defined. While I did not believe that the impossibility of defining an electron and the impossibility of defining God had much in common, yet it spared me the foolishness of setting up a definition of God of my own and then knocking it down again like an Aunt Sally, which is what so many atheists do. So I continued to think and think and think.

I am sure now that intelligent adolescents are bound to go through a period, often a very prolonged period, of scepticism, of seeking but not finding, sometimes of not even knowing what they seek. They engage, and should engage in philosophical inquiry about the nature of God and the absoluteness of right and wrong; about the technique of creation (and here I mean not just the creation of the universe but of man and his emotions, the power of communication, the origin of evil) and the possibility and meaning of immortality. I think it is very much to the credit of the Bishop of Woolwich that although, owing to the circumstances of his upbringing, he missed this early stage of his development, his mind was sufficiently elastic for him to be able in middle age to go through a period of deferred adolescence, and to share his thoughts with others. Each of us, at the time, of course thinks that his experience is unique; and few of us have the confidence to share it, except perhaps with a few contemporaries.

The real test comes when we have to teach our own children.

Philosophical questions do not worry the very small child. Facts and fancies they can absorb and they very often get them mixed up together, but philosophy makes no impression. I very well remember my elder daughter coming to me when she was about four years old and asking me what happens to you when you die. I answered her by reading to her one of Mrs Gatty's admirable *Parables from Nature* about the grub that turned into a dragonfly. I think I was trying to express the fact that we simply do not know what will happen and that instead of oblivion the sequel to death may be something that to us now is unimaginable. At least, that is the point of the story. But these stories are really for adults and not for children. I can read them now with considerable enjoyment. Jane listened politely and asked no more questions, but a day or so later I heard her put the same question to an older child from the nearby village. 'Oh', said the older child, 'they puts your body into a box and shoves it under the earth.' Jane was satisfied. This was fact, not philosophy.

The one thing I was determined to avoid in my children's training was the suggestion that anything must be swallowed whole, even the multiplication table. There was no danger that they would be required to subscribe to a creed. My husband and I had begun to attend Friends' Meetings when they were tiny, and Friends deliberately avoid creeds not only because words can mean different things to different people but because, as Eddington said: '... The spirit of seeking is still the prevailing one in our faith, which for that reason is not embodied in any creed or formula'.* Eddington went on to say that Quakerism in dispensing with creeds holds out a hand to the scientist; but, as Dr Harold Loukes has pointed out,† creedlessness chimes well also with modern educational thought, which offers opportunity of experience rather than formulae to be learnt by heart. Children need, above all, the experience of being loved unselfishly and of living in a home where gentle, kindly

---

* A. S. Eddington, *Science and the Unseen World* (London, 1929), p. 53.

† H. Loukes, *Friends and their Children* (London, 1958), p. 19.

behaviour is taken as much for granted as is honesty in thinking and truth in speech. We did not always achieve this, but I think it was clear to our children that we were trying to do so. In attempting to avoid any kind of religious education that a child may have to unlearn later we are, however, caught on the horns of a dilemma. Many an adult still thinks of God as a bearded old gentleman sitting on a throne beyond the clouds simply because that was the image of 'Our Father which art in Heaven' that they made for themselves as a child; yet as Sir Waiter Moberly has said in the *Crisis in the University*,* if you teach nothing about God then you do in fact teach in the most insidious way possible that he does not matter.

What we must be prepared to do is to discuss these matters with our children as they grow old enough to ask questions about them, and to show them – if indeed we can – that we ourselves have a living faith, based on our own experience. If we cannot do this, we can at least show that we believe the question important enough to be constantly thought about. They will grasp that, and will respect the honesty with which it is admitted. What is really paralysing is for us to stop thinking because we have made up our minds, and have persuaded ourselves that there is nothing more to discuss, even with ourselves. In scientific research this would be fatal to progress; and since scientific experience is part of my way of life it is certainly part also of my religious experience. In making this statement I am sure that I join issue with some of my scientific colleagues.

In the Lent Term of 1957, a group of seven scientists gave a course of lectures in the University Church, Cambridge, and these were subsequently published by the SCM Press under the title *Religion and the Scientists*. It is difficult to quote without some danger that the omission of context will alter the sense, but I think that the following rather long extracts do convey the sense of what Professor Mott said in the first lecture of the series:

---

* Sir W. Moberly, *Crisis in the University* (London, 1949).

The test of scientific truth, apart from its simplicity and elegance, is the success with which predictions can be made from it and technology built on it, and the wide consent which it subsequently commands ... Newton's laws, even if comprehended in the wider synthesis of Einstein, retain their beauty, their truth and their practical value, and all scientists are in agreement about this. I need hardly emphasise that religious thought does not advance in this way through experiment and widening agreement, and one may hope that it never will. The religious field of thought attracts – I speak personally – by its essential uncertainties, its fields in which there is no agreement, its glimpses of those problems in which despite their intellectual fascination no certainty is possible, except, if you so choose, through an act of faith. By these, I mean the most fundamental problems, the existence of God, the nature of God and his relation to man ... Although I agree that the contemplation of the greatest scientific generalisations may stand in a sense alongside religious experience, I would on the other hand even maintain that the day-to-day work of the scientist, the unravelling of one small detective story after another, the struggle with leaks, short circuits and mathematical tricks and the angling for financial support, gives him a training that is difficult to reconcile with religious thought. The trouble is that a scientist has such a definite idea of what he means by truth. Truth to him is what can be verified by experiment, what will serve as the basis of further experiments ... and what is agreed by scientists on both sides of the Iron Curtain. Religious truth is so demonstrably not of this nature. It is a pity that the literature of religion is not written still in Latin; to have a different word for it, such as *veritas*, would help ... I would say at once that one cannot on strictly scientific grounds object to the belief that, at the birth or death of a divine person, miraculous events occurred. The event was, in the belief of Christians, unique; science deals with events that can be repeated, so science cannot properly say anything about them. My own strong disinclination to accept these miracles is, I believe, based on aesthetic rather than scientific grounds; it springs from the feeling that they are unnecessary ... Nevertheless I would not wish any substantial change made in the historic Creed and form of service of any church which I attended ... 'Faith', said some anonymous schoolboy, 'is believing what you know ain't true', and may I describe one approach which a scientist may take to parts of organised worship as 'loving what you know ain't true' ... one can only know religion through the records of Christian and other religious experience; history and what has been written throughout time is of the essence of the matter.

I have quoted Professor Mott at considerable length because I am sure that what he says, so clearly, is also the opinion of a very large number of members of the Christian Church as well as of those who, agreeing with him, are as a consequence not willing to call themselves Christians.

In the second of the lectures referred to, Dr Mary Cartwright put the case for traditionalism very succinctly: 'So far as I can see, Christianity stands or falls by the resurrection more than by any other features of its teaching ...'.

Both these quotations are in line with the doctrinal basis of the World Council of Churches and of that now proposed for the British Council of Churches. Both, it seems to me, accept a definition of Christianity which not only imposes a strait-jacket on those who wish to call themselves Christians because they are humble learners in the school of Christ, but which, as I shall suggest later, has no place in the teachings of Jesus Christ himself: rather the reverse.

One of the first duties of a research scientist is to make a careful and critical study of all the work that has previously been done in the field in which he is interested, or at least of as much of it as is relevant. This is a practice that should not be set on one side when considering questions of faith or of ethics. I do not at all agree with Professor Mott when he suggests that truth in science and religious truth are demonstrably not of the same nature and that one ought to use different words to describe them. What we look at when we study the research work previously carried out in our own scientific field is the experience described in the literature. We then form our own judgement as to whether that experience was accurately recorded, whether the deductions from it were sound, whether the hypotheses put forward as a result were reasonable and whether the experiments thereafter carried out to test the hypotheses were well conceived. If we find that a particular scientist has a bee in his bonnet, that his experiments are so framed that they are bound to give him the answer he wants, then we reluctantly conclude that he has become obsessed by his theory and is not disinterested in his search for truth. We are, in short, looking for accuracy in the records and truth-seeking in the observer or experimenter. But let us make no mistake about this. Neither scientific inquiry, however honest, nor the technology built upon it, however successful, can *prove* in a logical and positive sense that the theory formulated is correct. This has frequently been emphasised but I shall quote two sources only: one a Friend, Thomas Kelly, who, in his *Reality of the Spiritual World*,* points out that one most frequent argument for the reality of God is

> the fact that lives that have experienced God as vividly real [in the sense described so well by Eddington in *Science and the Unseen World*] are new lives, transformed lives, integrated lives, souls newly sensitive to moral needs of men, newly dynamic in transforming city slums and eradicating war. By their fruits we know that they have been touched, not by vague fancies, by subjective, diaphanous visions, but by a real, living Power. The consequences of the experience are so real that they must have been released by a real cause, a real God, a real Spiritual Power energizing them.

If this argument were sound, there would be a close analogy with the acceptance of scientific truth because of the success of the technology built upon it. 'Look', says the modern scientist, 'it works!'

But, as Thomas Kelly points out, there is a logical fallacy in this pragmatic test: it is the Fallacy of Affirming the Consequent. I

* T. Kelly, *Reality of the Spiritual World* (Pendle Hill Pamphlet 21, 1942).

will put this fallacy in a form applicable to a scientific experiment: 'If this theory or hypothesis is true, it will have such-and-such a result. I do find this result and therefore the theory is true.' The fallacy is that you have not proved and cannot prove that your theory is *unique*, that it is the only possible theory that will lead to the result obtained. The negative statement is logically much sounder. 'If this theory or hypothesis is true, it will have such-and-such a result. I do not find this result and therefore the theory is not true.' This is a much less satisfying conclusion, however, because even the process of elimination cannot ensure that you have considered all the possible theories, and what we aim at is a unique solution. But, Thomas Kelly says

> if religious experience cannot be proved to be entirely reliable by the pragmatic argument, is religion alone in this respect? Far from it. I would remind you that the whole of experimental science which we revere today rests upon such argument, and faces the same predicament. Every scientific theory that is supported by experimental evidence rests upon the fallacy of affirming the consequent. The outcome is that the whole of scientific theory is probably only, not absolutely certain. But this fact has not paralysed science, which proceeds all undisturbed by the logical defect and, with open mind, lets down its faith upon its findings. For science rests upon faith, not upon certainty.

I said that I would quote a second source: it is Professor H. Bondi, who in his 1961 Granada Lecture said:

> One can never say that science proves anything. Empirical evidence can never prove a theory but can only disprove it ... Because scientists know they may be mistaken and are grateful to anyone who can demonstrate that they are mistaken, therefore people of different religions, different ideologies, different educational systems, different languages can work together happily in science because all agree on a yardstick

of experimental *disproof*, which is well-tailored to human gifts and limitations.*

This is not quite the whole truth, of course, as I am sure Professor Bondi knows. Scientists work happily together not merely because they have agreed upon a yardstick of experimental disproof but because, by and large, they have agreed to ignore the fallacy of affirming the consequent except that if they are particularly honest or particularly careful they qualify their conclusions by adding 'probably'. They do, however, pride themselves on being at any time ready and willing to withdraw or modify their theory in favour of a better. Nevertheless, there are many scientific experiments of which the underlying theory is so generally accepted that no one can really imagine that there can be any better theory until the day when new ideas or experiments suggest that the old theory is inadequate.

Nor is it really true that science is something rather unique in that scientists of different ideologies, etc., etc., can agree on it because they use the same yardstick to measure truth. This is so often said that a great many people believe it to be true and believe also that religious truth is demonstrably not of this nature'. I have been a crystallographer now for a matter of 42 years. During that time I have had many arguments with my fellow crystallographers about the interpretation of our various experiments. Scientists are very good at fighting rearguard battles, and they do not necessarily come to an agreement before they retire or die. There are some research problems on the solution of which we do agree. Why? Because we have had very similar training and have learned to interpret our experiments in accordance with that training. It is very seldom that an entirely new theory gains immediate acceptance, but when it does it is because those scientists whose training allows them to understand it find that it explains experiences not satisfactorily explained before, or that it ties up a number of loose

* H. Bondi, *Advancement of Science* (1961 Granada Lecture; Sept. 1962)

theoretical ends. If crystallographers on both sides of the Iron Curtain agree, *it is because they have had a similar scientific training*. Often they have used the same textbook, translated from one language into another. I believe (or at least I have been so informed, and I see no reason to disbelieve it, since the information has been found to be correct in other cases) that I have a small bank balance to my credit in the Soviet Union, royalties on the sales of one of my scientific books over there. Japanese, American, Russian, Chinese, Indian, European, South African, Egyptian and Israeli crystallographers have all had the same type of training; and so they agree, generally speaking although not entirely, on the basic interpretation of crystallographic experience. Even so, their field of knowledgeable agreement is small. They are agreed that X-rays are diffracted more effectively by electrons than by nuclei and that neutrons are scattered more effectively by nuclei than by electrons (at least in non-magnetic materials). But if you were to ask them what they mean by X-rays, neutrons, electrons and nuclei, I fear that their answers would be regarded as hopelessly inadequate by the spectroscopist or the nuclear physicist.

English, American and Russian chemists would know, or ought to know, the formula for anthraquinone. But confronted by an X-ray diffraction photograph of a crystal of anthraquinone they would not be able to identify it and it would take a considerable amount of training of a particular kind to enable them to do so. Unless they thought that kind of training important for their understanding of their own subject they just would not be willing to undergo it; they would take the crystallographer's word for it.

Scientists of different training do not in general agree except by the application of the pragmatic test; and even then usually in an uninformed way. Physicists have not tested the theory of evolution by the laborious process of elimination by disproof, nor do the majority of biochemists, I would hazard a guess, understand how Einstein's General Theory of Relativity has improved on Newton. Scientists of different disciplines do accept one another's word, because they in general recognise that they have had similar types of training and are seeking the truth in similar ways. If a new theory, say the Theory of Indeterminacy, gains general acceptance in one field, although probably not fully understood by a considerable number of working scientists even in that one field, then the most active and original research scientists in other fields will study it, to see whether it may have implications for them.

Now what is religious truth? If it is to be confined to philosophical conjecture as to the existence and attributes of God then it becomes an invention of the scholar and not a lifeline to the seeking soul. Religious philosophies do, of course, abound and some of them are mutually contradictory. There is a powerful analogy with the scientific disciplines in that those people who have had a similar training tend, on the whole, to agree among themselves. The majority of those brought up as Hindus remain Hindus although they may later become physicists, chemists, biologists, priests or road-sweepers. The majority of those brought up as Mohammedans remain Mohammedans and agree that there is one God and Mohammed is his prophet, whether they are Pakistanis or Arabs or Malayans; because they have had the same basic religious training. That is why the Catholics insist that the children, even of mixed marriages, shall be brought up as Catholics. And the doctrines of the Roman Church, although modified for the intellectual as distinct from those having more simple minds, transcend the differences between one race and another, because of this uniformity of training of young Catholics whether in Italy, Ireland or the United States of America.

I am not suggesting, of course, that there is a complete analogy between Buddhist and Moslem on the one hand and botanist and physiologist on the other; although if the various world religions are stripped of their overgrowth of superstition they may be thought to represent different facets of religious experience just as the different disciplines represent different facets of scientific experience. What I do say is that it is simply not true to imply that scientists agree among themselves in a way that religious devotees do not. Groups of scientists who have had similar scientific training agree; and groups of men of religion who have had similar religious training agree. Each may have minor differences of opinion or method even within their own group. What is true, however, is that generally speaking scientists in one discipline accept, usually in an uninformed way, the conclusions of the specialists in other disciplines; whereas each religious group tends to think of others as mistaken. This attitude, however, is changing; there is more tolerance between one Christian sect and another, more liberalisation of all religions and more willingness to seek what is common to all. It is only natural, perhaps, that we should be at different stages of development in different fields of thought, even as individuals, and certainly as groups.

So far I have been comparing religious doctrines; when it comes to religious experience there is a wide variety within each world religion and a clearer recognition of what is basic to them all, or nearly all. The varieties have been most graphically described by William James in his famous 1901–2 Gifford Lectures* which he himself described, not quite accurately, as 'all facts and no philosophy'. Stripped to the limit, what all religions have in common is a sense of need, an uneasiness, a sense that there is something wrong about us, and a sense that our need may be met, that we can be saved from the wrongness by a proper connection with a power beyond ourselves, a 'higher' power or powers. We recognise goodness (even when we cannot define it and do not all agree about its manifestations) as something desirable, and we apply to any religion the pragmatic test: does it produce good men and women? Maybe we know best the saints of our own denomination, but there can be few Hindus or Christians who would not admit that both Francis of Assisi and Mahatma Gandhi had qualities of saintliness.

In the widest sense (and it is in this wide sense that Friends understand religion) religious experience includes all experience,

* W. James, *The Varieties of Religious Experience* (Fontana Edition, 1960).

and it is in this sense that Harold Loukes describes the content of religious education as

> the same field of experience as that of any education, but observed with a constant readiness to see it whole, and refuse to be content with parts: to love and enjoy flowers as well as to analyse and classify them; to enter imaginatively upon history as well as to observe the chain of causation; to seek to understand the human plight of other races, instead of viewing geography as only the interaction of climate and structure upon vegetable, animal and human species. Any education which seeks to penetrate into the significance of experience is, in a measure, religious education.

It is obviously not the whole of religious education and Harold Loukes adds to it in another essay as follows:

> Although we can ... convey nothing to our children of religious concepts, this does not mean that we cannot offer them the beginnings of religious experience. This ... is the beginning of Christian education: the offering of the experience of being loved. But this, it may be objected, is not Christian experience: it is merely the common experience of humanity. Any baby can experience this: Muslim babies, Hindu and Buddhist babies, agnostic babies, Marxist babies. What is there Christian about this? Such an argument mistakes the whole nature of Christianity, which is not a denial of human nature, but an affirmation of it ... There are other wordless lessons that our children will learn so long as we do not prevent them, experience lying to their hand that is in very truth the experience of God: the sense of wonder and mystery; the sense of dependence, the discovery of compassion – all these things lie about us in our infancy ... *

My argument, then, is that religious truth is known to us through religious experience, our own and other people's, and that reli-

* H. Loukes, *Readiness for Religion* (Pendle Hill Pamphlet 126, 1963).

gious experience includes scientific experience while going far beyond it. Not all Christians would admit this. In a course of lectures given at Woodbrooke College, Birmingham, some 20 years ago and published by the SCM Press under the title *The Tyranny of Mathematics*, a well-known Quaker headmaster, Geoffrey Hoyland, claimed in a very witty and persuasive way that the scientific method is to pull the thing you are studying to pieces, look at the bits and then put it together again. Since from the scientific point of view, he said, the whole is invariably equal to the sum of the parts, you will then have found out the scientific truth about it. We would have no quarrel with this method, said Geoffrey Hoyland, if the modern world did not elevate science into the role of a god, and if that all-powerful, all-dominating influence, the scientific method, did not sometimes lead us away from the truth, as it does, he claimed, when we try to analyse the beauty and poetry of Rembrandt's 'Man in a Gold Helmet', Salisbury Cathedral, Keats's 'Ode to a Nightingale' or the B-Minor Mass. Geoffrey Hoyland has said:

> In the sphere of the Christian faith the scientific method, with all its apparatus of criticism, has laid siege to the Gospels and has shredded them to provide fodder for its microscope. And, more serious still, many of the critics, when they have completed their analysis, insist that the Calculus principle must apply equally to the subsequent synthesis; they will not admit that the whole can be anything more, or anything different from, the mathematical sum of such parts as they have left undestroyed. Is this sound criticism – is it getting at the truth, or is it leading us away from it as it did in the case of Rembrandt and the Ode?

But this is really a caricature of science; it is quite true that in crystallography, the science that I know best, when we take, say, a single crystal of a chemical compound and analyse it by means of X-ray diffraction techniques we are, in a sense, pulling it to pieces and looking at the bits. We obtain diffraction patterns consisting of spots or lines on a photographic film or a series of peaks recorded by a counter method. We measure these, process and synthesise our data by means of a computer and the result with some intelligent interpretation tells us how the atoms are arranged in the unit of the periodically repeating pattern which gives that kind of crystal its special properties. Whether we examine a grain of salt, a diamond, the virus of polio or a crystal of haemoglobin the method is essentially the same. But the acquisition of fresh knowledge by such an analysis does not remove all the beauty from a crystal or make us hard and insensitive. On the contrary, it adds to our appreciation of it, when we realise that the outward form is the consequence of an inward regularity and periodicity of a kind which is unique to every species examined. We wonder, too, at the beauty of many of the diffraction patterns recorded, and such patterns have found a place in a number of books on art. If it is true, as I think it is, that emotion must not influence scientific observation, it does not at all follow that emotion is eliminated from the scientific life. In 1933, in an article on *The Struggle for Intellectual Integrity*, P. W. Bridgman, the well-known physicist, wrote:

> Once the scientist has started living the life of scientific honesty … he finds growing within him the realisation that he is in possession of something more than merely a tool by which he may get the right answers. The ideal of intellectual honesty comes to make a strong emotional appeal … *

Appreciation of beauty and order, and appreciation of intellectual honesty move us towards the truth, not away from it.

When it comes to the dissection of a living organism to find out what makes it work, it is quite true, of course, that if the surgeon finds that his patient is dead he cannot bring him to life again by just sewing up the parts; 'the silent thing on the table is not a man'.

---

* P. W. Bridgman, 'The struggle for intellectual integrity', *Harper's Monthly Magazine* (1933).

But it is a knowledge of anatomy that enables the surgeon to save life by the operation that is carried out in time, and it is, or should be, his concern to relieve suffering humanity that provides the incentive for the arduous training necessary to direct his skill.

Nor were such masterpieces as the 'Man in a Gold Helmet', Salisbury Cathedral, the 'Ode to a Nightingale' or the B-Minor Mass created as a whole. They were built up by specialists who were masters of their craft, detail by detail, and we admire and appreciate them no less if we understand the work that went into them. The mistake would be not to see them as a whole *because* we were interested only in the detail.

Geoffrey Hoyland comes round to this when he concludes that it is the service and not the domination of the scientific method that the world needs. But is it a fact, as he implies, that the scientific method, with all its apparatus of criticism, when applied to the Gospels leads away from the truth instead of towards it? My answer to that is that it should not do so if we are really seeking truth and not what Sylvanus P. Thompson called 'adventitious aids to truth'. Jesus himself rebuked those people who wanted some miraculous proof of the authority of his teaching. 'Except ye see signs and wonders', he said, 'ye will not believe', and 'Can ye not discern the signs of the times?'

Sylvanus P. Thompson makes this comment on the prevailing attitude at that time:

> Yet everywhere, even in the case of false prophets and false teachers, signs and miracles were regarded as somehow proving the truth of the doctrines taught ... On the modern mind the effect is just the opposite. If any modern teacher of religion or morals would propose to establish the truth of his teachings by showing some unaccountable marvel – by working a miracle in fact – the sincerity of his teaching would be at once discredited. We should rank him straightway as an impostor. Our belief in the teachings of Jesus Christ ought to be held by us *because we are convinced of their inherent truth*, not because he is said to have worked miracles ... To see the narratives of miracles in

> their true proportions we have got first to appreciate, and enter into, that Oriental state of mind which, with perfect honesty and sincerity, values the adoring legend because it is adoring, more than the naked truth because it is true.*

The *disbelief* in the sincerity of any modern teacher who would attempt to prove his credentials by working miracles is one of those mutations of religious thinking to which I shall refer later; and it is to modern science that religion owes this advance, for advance I am sure that it is. May I quote again from Sylvanus P. Thompson:

> The Apostolic advice: 'Prove (test) all things; hold fast that which is good', should be an incitement to follow the quest. In that pursuit our business is to demand evidence, to evaluate its weight, and to be tenacious of that which has been found to be demonstrably true... To refuse to admit Truth to scrutiny, lest it should fail to meet the test, is cowardice, not faith.

My own objection, now as then, to the doctrine of salvation to which I was required to subscribe as a child, is that it demands the uncritical acceptance that we would rightly refuse to give in the case of a scientific theory. That does not, of itself, prove that it is untrue. In a most persuasive essay entitled *On obstinacy in belief*, C. S. Lewis says:

> There are times when we can do all that a fellow creature needs if only he will trust us. In getting a dog out of a trap, in extracting a thorn from a child's finger, in teaching a boy to swim or in rescuing one who can't, in getting a beginner over a nasty place on a mountain, the one fatal obstacle may be their distrust. We are asking them to trust us in the teeth of their senses, their imagination and their intelligence. We ask them to believe that what is painful will relieve their pain and that what looks dangerous is their only safety. We ask them to accept

* S. P. Thompson, *The Quest for Truth* (Swarthmore Lecture, 1915).

> apparent impossibilities: that moving the paw farther back into the trap is the way to get it out–that hurting the finger very much more will stop the finger hurting – that water which is obviously permeable will resist and support the body – that holding on to the only support within reach is not the way to avoid sinking – that to go higher and on to a more exposed ledge is the way not to fall. To support all these *incredibilia* we can rely only on the other party's confidence in us – a confidence certainly not based on demonstration, admittedly shot through with emotion, and perhaps, if we are strangers, resting on nothing but such assurance as the look on our face and the tone of our voice can supply, or even, for the dog, on our smell. Sometimes, because of their unbelief, we can do no mighty works. But if we succeed, we do so because they have maintained their faith in us against apparently contrary evidence. No one blames us for demanding such faith. No one blames them for giving it. No one says afterwards what an unintelligent dog or child or boy that must have been to trust us. If the young mountaineer were a scientist, it would not be held against him, when he came up for a fellowship, that he had once departed from Clifford's rule of evidence by entertaining a belief with greater strength than the evidence legally obliged him to do. Now to accept the Christian propositions is *ipso facto* to believe that we are to God, always, as that dog or child or bather or mountain climber was to us, only very much more so.*

It is fair to point out, however, that every example given by C. S. Lewis in this passage involves not so much faith, trust or belief, in the philosophic sense of 'acceptance', but *action*: to move, or allow someone to move the paw further back into the trap, to hold the finger still while the thorn is extracted, to place one's body into the position in which the water will support it, to remain quiet and passive while being rescued, to climb to the higher ledge. And it is also possible 'to believe that we are to God, always, as that

* C. S. Lewis, 'On obstinacy in belief'. Reprinted in *They Asked for a Paper* (London, 1962).

dog or child or bather or mountain climber was to us, only very much more so', *without* accepting the doctrines of the Incarnation or the Resurrection, which the context shows was what C. S. Lewis meant by 'the Christian propositions'.

Man's sense of need and the consciousness that there is a power beyond his own that can help him is, as I have already said, basic to almost every religion, however much it may be overlaid by other accessory doctrines. My own disinclination to accept the supernatural doctrines traditionally associated with Christ's person arises not so much from scientific scepticism concerning miracles as such, nor from the aesthetic sense that Professor Mott refers to, that they are unnecessary, but from my inward and considered conviction that they are contrary to the whole tenor of Jesus's own approach to his message. He must quickly have realised that he had tremendous power to influence men and women. He could have been a political leader. He refused. He was tempted to use his power to force people to believe in him and so to accept his teaching. It must have been a temptation that came again and again when he was dealing with a people so credulous and when they expected signs and wonders of him. Yet again and again he insisted that he did nothing that other people could not do and that when they came to him for healing and healing occurred, it was by their own faith, or the faith of those who loved them. This part of the record is consistent with his honesty and integrity; it accords with our whole picture of the man himself; it fits in well with his teaching that thought, word and deed must be clean and upright and in harmony one with another; that those who love God must also love and desire the good of others, even their enemies; and that nothing must be done for self-aggrandisement or for self-justification. That such a life should be associated with an event so miraculously unique as the resurrection of Christ's body from the tomb in which it had been laid, so that he could eat and drink again with his disciples, a literal coming to life again in such a way that doubting

Thomas *had* to believe, would be a glaring contradiction, not the crowning proof of a religious truth. Or so it seems to me.

There are of course many Christians who also reject the crude and materialistic doctrine of the resurrection of Christ's body, as well as of their own. But simple folk are still taught the crude story as essential to Christianity. Some of the early Fathers used to support the doctrine of the resurrection of the body with arguments that must once have seemed reasonable to them although now they seem fantastic. Saint Jerome asked 'If the dead be not raised, how could the damned, after the Judgement, gnash their teeth in hell?'. This reminds us of the wicked Irishman who, threatened with hell by his priest, retorted that he had no teeth left to gnash; only to be told 'Never comfort yourself with that, Patrick. Teeth will be provided.' But St Paul, who was one of the strongest upholders of the doctrine of the resurrection, was very clear that he was talking of a spiritual body and not a natural one. Translated into language more familiar to our way of thinking, he was arguing for the continued existence of the individual personality or soul after it is freed from its human encasement. If this is so it will, sooner or later, become the experience of each one of us. Until then we cannot know, any more than Michael Faraday could know that men would one day be able to talk to one another over distances of thousands of miles. Does it make any difference to our life whether we believe this or not?

What seems to have changed is our personal attitude to the idea of immortality. Many of us simply could not now agree with the dictum that 'If in this life only we have hope in Christ, we are of all men most miserable'. On the contrary, we would say that if even in this life only we may have the opportunity and experience of serving God and of being used by him then we are of all men most fortunate. The sequel, if any, we are content to leave. In this day-to-day service we meet the equivalent of the working scientist's struggle with leaks and short circuits: we are not concerned most of the time with major problems of the meaning of life, we are simply asking God's help in our constant battle with a quick temper, or a poor digestion, or a failing memory.

Does it surprise you that I should now be talking about 'being used by God' and 'asking God's help' when at an earlier stage of this lecture I was questioning whether one could honestly believe in a God who could not be defined? Well, I can only explain myself by saying how totally my experience differs from that of Professor Mott when he said that 'religious thought does not advance ... through experiment and widening agreement'. I am sure that it does so advance both in respect of our personal experience and in respect of human thought generally. Perhaps I may deal with the last first. Here I quote from William James (Lecture 14):

> To the extent of disbelieving peremptorily in certain types of deity, I frankly confess that we must be theologians. If disbeliefs can be said to constitute a theology, then ... prejudices, instincts and commonsense ... make theological partisans of us whenever they make certain beliefs abhorrent. But such commonsense prejudices and instincts are themselves the fruit of an empirical revolution. Nothing is more striking than the secular alteration that goes on in the moral and religious tone of men, as their insight into nature and their social arrangements progressively develop ... Today a deity who should require bleeding sacrifices to placate him would be too sanguinary to be taken seriously. Even if powerful historical credentials were put forward in his favor, we would not look at them. Once, on the contrary, his cruel appetites were of themselves credentials. Few historic changes are more curious than these mutations of theological opinion.

William James went on to say that when men thought of God as a sovereign ruler, a dose of cruelty and arbitrariness was positively required of him; and it is not much more than two centuries since an otherwise lovable New England Puritan, Jonathan Edwards, described his conviction that God dealt out salvation and damnation quite arbitrarily to selected individuals as a 'delightful conviction', as of a doctrine 'exceeding pleasant, bright and sweet'. Our present rejection of such irrational and mean behaviour comes, said William James, from

> the voice of human experience within us, judging and condemning all gods that stand athwart the pathway along which it feels itself to be advancing. Experience, if we take it in the largest sense, is thus the parent of those disbeliefs which ... were inconsistent with the experiential method.

But of course new ideas do not come simultaneously to all men at the same time. What does happen is that one man challenges tradition and his challenge strikes a sympathetic chord in other hearts so that his ideas spread. Jesus himself threw out a tremendous challenge to the theologians of his day – the scribes and pharisees. He challenged not so much their teaching as their practice; but the God whom he revealed by parables and simple analogies to his followers was really a very different God from that of the scribes and pharisees. And what he asked of his twelve disciples was what can only be described as an experiment. They were to go out as itinerant preachers, with no money, no change of clothing, no planned lodging for the night. They were to bring healing, as he brought it, to those who had faith. They were not to rehearse in advance what they should say, even when brought before governors or kings. God would look after them, and speak through them. God was their Father and loved them; and men would know that they were Jesus's friends and disciples if they loved one another.

What seems to me to be so tragic is that because of their passion for signs and wonders and their unwillingness to experiment with love and trust instead of hatred and suspicion in their dealings with one another, men have ignored this simple teaching of Jesus and have obscured his message by weaving around him a web of tradition. To the medieval theologian this earth *had* to be the centre of the universe because it was here that Jesus came. To suggest that the earth was just a minor planet in our universe was therefore heresy. And the trouble is that this attitude still prevails. Last year I was asked by sixth formers in Birmingham whether it would make any difference to the *truth of Christianity* if spacemen found living beings on other planets. It is a variant of the same idea. The trouble about the attitude of those who, like Professor Mott, have an affection for the tradition even though they know it is not true, is that while this affection may hold the Christian Church together for one generation, it will not continue to do so. Meanwhile the archaic language of the Creeds misleads and confuses those who hear it. Either they accept it at its face value, or they interpret it in their own way, or they reject both it and Christianity as being out-of-date. And yet the Church could have a message that the world greatly needs if it taught what Jesus taught, instead of adding or substituting dogmas that have no meaning for the honest seeker.

In Volume 1 of the *History of the People called Quakers* (1789) John Gough writes: 'We consider ... that religion most worthy of our study and pursuit which mends the heart and regulates the life and manners, not that which only fills the head with a notional apprehension of divine things.'

Early Friends had a very special use for the word 'notion', which I can best make clear by an example. When my children were small we were once talking to an elderly relative and she said 'I believe that some of the other planets are inhabited by living beings. I believe it but I can't prove it.' Then turning to the children she added 'That is faith'. But I could not let that statement go by unchallenged and I had to say: 'Oh no; that is not at all what is meant by "faith". That is what Friends call a "notion". A notion is an assertion that is unrelated to our experience, that makes no difference to our way of life. If, however, someone believed so strongly in the possibility of life on other planets and thought it so important that they devoted all their money or time or skill to the promotion of interplanetary travel then to them it would no longer be a notion, it would have become a hypothesis to be tested by experiment, even though they might be well aware that they themselves would probably not live to see the answer.

We know what we mean by a hypothesis in scientific research, and if our career consists of or includes the carrying out of such research then we know that a hypothesis can come to us in one of several ways. It may be suggested by our reading of the scientific literature; it may be suggested to us, directly or indirectly, by our discussion with other scientists in our own or allied fields; it may be the result of deduction from our own previous experiments or mathematical analysis; or, more rarely perhaps, it may be the result of scientific intuition: a 'hunch'.

To each one of us the whole of life is an experiment. We have not gone this way before and most of us would deny that we are automata. Undoubtedly the kind of life we live and our approach to it are conditioned to a greater or less extent by our background, environment and training. So it is in science. But what we do not always realise is that in the experiment of life we cannot proceed without *some* conscious or concealed hypothesis. We may think it does not matter how we behave (although there are few people who do not have a sense of 'ought'); if so, we probably live merely to please ourselves. We do the same if we believe that there is no aim higher than our own personal happiness. We may believe that what is good for Britain is good for the world; if so, we pursue nationalistic aims with a clear conscience. We may believe that other countries, especially those whose ideology we dislike, understand nothing but force. If so, we build up our armaments and adopt a bullying or threatening attitude. In some cases our belief amounts to a conviction rather than a hypothesis, and such a conviction can be based on reasoning or emotion rather than past experience. If it is based on reasoning we are more likely, I suggest, to be able to admit that we are wrong than if it is based on emotion. But sometimes we genuinely want to do the 'right thing' and do not know what it is. We want, perhaps, to know how best to deal with criminals or young psychopaths or a bad-tempered elderly relative or a very difficult child, especially when we realise only too well that we ourselves are not saints. Scientific knowledge is clearly not enough here; there is not enough of it, for one thing (we could certainly do with more); and the problems are ones of personal relationships, not just of treatment.

Much of our penal treatment in this and other countries has been, and still is, frankly experimental and is based on the presumption, the hypothesis, that a certain percentage of criminals can become good citizens if properly re-educated, re-formed. It is therefore partly punitive, partly preventative, partly reformative; and the balance of each of these and the techniques vary in different institutions and from time to time. Anyone who has to deal with those who work in the penal service knows very well that many of these men and women are inspired by high religious motives; and that the relatively humane treatment of delinquents begun in a most tentative way about 150 years ago through the prodding of people like John Howard and Elizabeth Fry has made its way because by-and-large it works.

Nearly 170 years ago William Tuke, a Yorkshire Quaker, founded the first mental hospital in England where mental disorders were regarded as an illness requiring both scientific treatment and deep personal understanding, the aim being to restore the patient to physical, mental *and spiritual* health. William Tuke believed that this could be done because he believed in a God of love and, as Quakers do, in 'that of God' in every man; and although he had no *proof* that mental disorders could be cured by a combination of kindness, sympathy and scientific treatment, he believed it worth while to make the experiment. It was not just a medical experiment, it was a religious experiment and it worked so well that it had a profound influence on public opinion and resulted in fundamental changes in the laws relating to the treatment of mental illness.

Now I maintain that by this kind of experiment and widening experience, religious thought in the best sense of the word does advance, and one may hope that it will go on doing so. But when it comes to our own personal lives, to our own bad temper or lack of patience with a trying old parent, what can we do? We may have tried and failed again and again. One thing we can do is to find out why others have succeeded better and to try their method. 'We cannot afford to throw away the experience of all the saints throughout the centuries.' What *all* religions have in common is this sense of need, and the reaching out to a higher power for help. Jesus was tempted, and he turned again and again in prayer to God. I do not think it is necessary to have a complete religious philosophy before we begin to ask for help to live a better life, especially if we are at our wit's end as to how to do it on our own. And the man or woman who is satisfied with themselves needs help even more than the one who can only pray 'God be merciful to me, a sinner'.

What I am trying to say, not very successfully perhaps, is that since we cannot live at all without *some* hypothesis a good hypothesis for living is more important than the intellectual acceptance of any idea which does *not* lead to good action. 'This people honoureth me with their lips, but their heart is far from me.' 'Why call ye me Lord, Lord and do not the things which I say?' If our study of the lives that we most admire leads us to conclude that they are good because they are convinced that they do experience real communion with God and receive help in doing so, then it seems only reasonable to believe, even though we are not yet convinced by our own experience, that this is possible and it is possible for us as well as for them.

Some of us are like a blind child who has never seen the sun. When we can see the sun, to say 'I believe in the sun' is unnecessary and almost foolish. But how does one begin to explain the sun to a child blind from birth? I don't know, but I imagine we might begin with analogies. Any one can feel the materials from which an ordinary coal fire, a gas fire, an electric radiator, a hot pipe are made. They are all different. Yet they all have one quality in common. At times they give out heat, so much heat that they can warm us without our touching them, although we know that they are not far away and that their heat can be controlled. The blind child knows what the shape of a sphere is because he has felt balls of different sizes. None of these is hot. He has experienced the warmth of a hot summer's afternoon and the cool of the evening or of a cold house on a wintry day.

When we first explain to him that this heat comes from a very hot sphere so far away that no one could ever travel to it even if it were cool enough, no one could ever touch it to find out what it feels like, he might at first say 'I don't believe it'. We would tell him too that it is the movement of this very hot sun relative to ourselves that explains the difference between the warmth of day and night, summer and winter, and that it gets cool even during a summer's day if something comes between the sun and ourselves. We would have tried to explain 'light' and 'seeing' and we would tell him that the sun gives not only heat but light and that it is the light and the absence of light that make the main difference between the day, when the birds begin to sing and people get up from bed, and the night, when the birds stop singing and people go to sleep. As the child grew older and more able to reason, the explanations could become more detailed, but more important still, he would find out for himself that this hypothesis 'the sun' that he could not see and could still only imagine in terms of analogies explained a great deal of his own experience as well as the behaviour of other people. He would realise, too, that there was no other explanation equally good. And at last he would be able to say 'I believe in the sun' with the conviction that came from reasoning and experience combined. He would not, perhaps, be able to enter fully into the experience of others, of people who could see as well as feel; and he would owe a good deal to those whose experience and patient affection had made him realise that this might be a reasonable hypothesis after all. But it would be his own reasoning, his own experience that had convinced him, and it might well be that his increased sensitivity to feeling would give him some experience that people with sight but less feeling could not share. The blind child who had accepted the idea of 'the sun' only on the authority of his teacher, could still enjoy the warmth of a summer's afternoon; but I venture to think that the sceptic might have gained something in understanding; and that his 'Now *I believe* in the sun' would have more meaning than if he had accepted the proposition unthinkingly. Yet whether he accepted the idea unsceptically, or accepted it as a working hypothesis to be tested against the facts, or was convinced of its truth by his own experience, he would still have had the experience. The sun would have been there all the time.

For one man the experience of God comes naturally; he grows up into it. To another it comes suddenly, although it may be that his heart and mind are already subconsciously prepared, as were Paul's by his witnessing the death of Stephen. Another still must seek God and cling to the glimpses he has: another by practising the presence of God, as St Lawrence did, can find him everywhere. Does this seem unfair? So does a great deal of life: the fact, if you like, that one man is born with temptations of which another knows nothing. But my own approach to a religious philosophy is similar to that of William James when he wrote:

> I *can*, of course, put myself into the sectarian scientist's attitude, and imagine vividly that the world of sensations and of scientific laws and objects may be all. But whenever I do this, I hear that inward monitor ... whispering the word 'bosh!' ... The total expression of human experience, as I view it objectively, invincibly urges me beyond the narrow 'scientific' bounds.

I have also heard that inward monitor saying 'bosh!' to atheism just as clearly as, in my early adolescence, I heard it say 'That can't be so!' to the fundamentalist attempt to put God into a strait-jacket. And I can add my witness to that of those who have proved the truth of Jesus's words 'Ask, and it shall be given you; seek, and ye shall find; knock, and it shall be opened unto you'. We do find help particularly in times of special need. God was most real to me when I was in prison, perhaps because so many of my friends were praying for me. We may also experience a 'dark night of the soul', which becomes more endurable when we realise that others have had the same experience and come through it. Our intellect may be perplexed and baffled by the problems of evil, of disease, of natural calamity, of undeserved suffering, and yet in spite of this we may have a sense of God's love that changes our belief in God from a hypothesis to a conviction, and enables us to tackle the problems of the sick world as well as our day-to-day relationships with others and our own bad temper.

It seems to shock some people that it should be possible to equate the words 'I believe ... ' with a hypothesis, or to suggest that one may live and act experimentally. Of course, we all do live a life of faith every day. When we get on a train or board an aeroplane we are putting our lives in the hands of the driver or the pilot. We are trusting them, although we know nothing of them personally. Our justification is that others have safely done the same before us, although the further back we go the more adventurous the experiment must have been. And God does call some of us to lives of adventurous experiment, for which there may be no precedent, at least as far as we ourselves are concerned.

I have sometimes been asked what were my reasons for deciding on that refusal to register for war duties that sent me to Holloway Jail 22 years ago. I can only answer that my reason told me that I was a fool, that I was risking my job and my career, that an isolated example could do no good, that it was a futile gesture since even if I did register my three small children would exempt me. But reason was fighting a losing battle. I had wrestled in prayer and I knew beyond all doubt that I *must* refuse to register, that those who believed that war was the wrong way to fight evil must stand out against it however much they stood alone, and that I and mine must take the consequences. The 'and mine' made it more difficult, but I question whether children ever really suffer loss in the long run through having parents who are willing to stand by principles; many a soldier had to leave his family and thought it his duty to do so. When you have to make a vital decision about behaviour, you cannot sit on the fence. To decide to do nothing is still a decision, and it means that you remain on the station platform or the airstrip when the train or plane has left.

If we knew all the answers there would be no point in carrying out scientific research. Because we do not, it is stimulating, exciting, challenging. So too is the Christian life, lived experimentally. If we knew all the answers it would not be nearly such fun.

THE VALUE OF LIFE

❀❀
❀

Baroness Mary Warnock
DBE FBA

THE THIRTIETH
ARTHUR STANLEY EDDINGTON
MEMORIAL LECTURE

26 January 1989

The Value of Life

MY LECTURE TONIGHT HAS an extremely wide title and I want straight away to narrow it. I shall be concerned with *human* life, not the life of plants, or animals other than humans. But there is a significance in this narrowing: in itself it reflects a value-judgment. It is not that there is *no* value to be ascribed to life in general. It is only that we give preference to humans over all other forms of life. Simply being a member of the species *homo sapiens* carries with it a value not generally accorded to members of other species.

Now it has been argued that to assert the priority of the human is to express a prejudice no more acceptable than the prejudice which places men above women, or white men above black. 'Speciesism' is supposed to be the name of this prejudice; and I cannot simply assume that we are right to place a higher value on human than on other forms of life without at least briefly considering the arguments of those who would accuse me of 'speciesism'.

One of the exponents of the view that the lives of at least some other animals are as valuable as the lives of humans is the utilitarian philosopher, Peter Singer, of Monash University. In the prologue to a collection of essays called *In Defence of Animals*, he argues that insofar as animals are capable of suffering they must *all*, not only humans, be taken into account in the calculus of pains and pleasures necessary for determining whether a kind of action is morally right or wrong. In this he is following the first Utilitarian, Jeremy Bentham, who said "The question is not Can they Reason? nor Can they talk? But Can they suffer?" and he went on "It may one day come to be recognised that the number of legs, the villosity of the skin or the termination of the *os sacrum* are reasons equally insufficient for abandoning a sensitive creature to the caprice of a tormentor "But Bentham was not prepared to follow the use of this criterion (the criterion of suffering) to its conclusion. It would have led him too far from what was legislatively practical; and he was above all a legislator not a philosopher. However J. S. Mill, though he too regarded the criterion of right as the balance of pleasure over pain, introduced a distinction between different kinds of pleasure and gave superior weight to the pleasures of the civilised over all others. Children and animals could feel only low-grade pleasures, and therefore could count for less than adult humans in the calculus.

Singer, on the other hand, argues that we should not give humans, adult or otherwise, priority, since there are no characteristics which definitively mark off human from other animals. If, following Aristotle, we seek to define humans as uniquely rational, then at once we get into difficulties about the meaning of 'rationality'. There are all kinds of mental activities, including kinds of calculations, that animals other than men undertake. The more we know of the capacities of dolphins, chimpanzees, even rats, the less we can believe in a total discontinuity between humans and other animals. Moreover (and Singer relies mostly on this argument) we do not know what to say about those humans who, though manifestly members of the species, are not able to do the things characterised as rational, and will never be able to do so. Singer thinks we ought to prefer the interests of the 'higher' animals to those of defective humans, on the grounds of their superior mental activities.

Now, unsurprisingly, Singer wants us to be vegetarians; but, rather oddly, he advocates this on the grounds that killing animals causes them to suffer. If we could find a way to kill them humanely, without any suffering, it seems that he would allow us to eat them. But he never faces the question whether, at a pinch, we ought to prefer to eat a defective human than an animal; nor whether, if we could be sure of causing no pain, we should be entitled to kill and eat a human, if we were hungry, or use him for our own ends, as at present we use horses or laboratory rats.

The fact of the matter is that Utilitarianism, in its concentration upon the criterion of pain and pleasure, suffering and happiness, cannot take into account the quite different sources of the obligation, which binds us to fellow-humans in a way that we are not bound to members of other species. We must not kill humans for the table, or use them for laboratory experiments, not because this would cause pain but because to humans we owe *justice*; and justice demands that they be treated by us as potentially, at least, our equals, equally capable of choosing and thinking and deciding for themselves, and of understanding their own position in the world. And, by courtesy, we allow such abilities to people whom we know not, in practice, to have them, simply on the grounds of their membership of that species which *does* as a whole harm these powers ascribed to it. We extend our concept of humanity to *all* members of the species, even though they may in fact be disabled or incompetent. We take on the helpless, or at least morality demands that we do, simply because they are our own.

'Speciesism' then is not a kind of prejudice. It is the necessary attitude of one species to another. It is true that our species alone has the power to think of the universe as a whole, and to consider other species and other lives. But that shared power is precisely what makes humans more important than other animals. It is humans, and they alone, who can be moral agents, who can make moral choices and judge themselves to be doing well or ill; who can see that other humans must be treated as free, choice-making individuals, since they are so themselves. Though we must treat other animals without cruelty we cannot admit them into the community of the moral. For because they would be incapable of recognising that they had been so admitted, or of admitting us into their community.

Granted therefore that we should think that harming humans matters more than harming other animals, that killing them requires a different justification, are we therefore committed to the view that all *human* life is of equal value? This is the question I want to raise this evening. I will start with a specific case.

Let us consider the notorious case of Baby Doe, born in Bloomington, Indiana, in 1982. It is a long sad story, with important implications. The baby was found at birth to have Down's Syndrome, and, in addition, to have a malformation of the oesophagus such that normal feeding by mouth would be impossible without immediate surgery. The surgery could not be carried out at the hospital where he was born, so the question was whether to transfer him to another hospital and operate, or leave him where he was, cared for, but certain to die. There was disagreement among the paediatricians concerned with the case; but when Baby Doe's parents were consulted, they decided that they did not want the operation performed, and they signed a statement to that effect.

However the hospital administration contacted the local County Court and an immediate hearing was set up. At this hearing the judge ruled that the parents had a right to choose between two medically recommended courses of action, and that there was no case to answer. Various attempts by the hospital administration and others to get this ruling put aside by a higher court failed, and while the attempts were still being made, the baby died. Public reaction was violent, and the most remarkable outcome was a memorandum from President Reagan, ending with the words "I support Federal laws prohibiting discrimination against the handicapped, and remain determined that such laws shall be rigorously enforced". The immediate outcome of this memorandum was the publication of posters to be placed in the intensive care units of all federally funded hospitals, indicating that funds would be withdrawn if handicapped babies were not so treated as to keep them alive.

Paediatricians now found themselves in obvious difficulties. The so-called 'Baby Doe Guidelines' were not guidelines but absolutely definite instructions with penalties attached. But were they really meant to apply to *all* babies, even if born with only fragments of a brain, or with cranial haemorrhage so severe as to make it impossible that the baby could ever achieve any cognitive abilities? By means of modern technology such babies could be kept alive, for months, even for years. But almost every doctor would think it right to let them die. The principle incorporated in the guide-lines and now about to be enforced by law was the principle that *all human life is of equal value*.

Can we actually subscribe to this principle? If we did value all human life equally, then it would be as important to preserve the life of a two-cell pre-embryo as of a fully grown adult; it would be as important to continue in life a terminally ill unconscious man of eighty as a young, healthy child in need of surgery after an accident. In such cases we take two factors into account. First we consider whether the creature *has a life that it is leading*; secondly we take in account the *quality* of that life. These two factors are interdependent.

Some people, when faced with the realisation that sometimes choices of this kind have to be made in practice, wish to draw a distinction between causing death or killing, which they would maintain is always wrong, and allowing to die, which they hold might be sometimes inevitable, and therefore justified. There is a considerable literature concerned with this distinction, but I have not time to explore it fully. I would argue however, that to maintain the distinction between causing death and allowing to die is to make a fundamental mistake about the nature of causation. People are often unthinkingly inclined to hold that a cause must be some kind of 'active agent'. Thus they suppose that the typical causal agency is that of a brick hurled through a window, which causes the glass to shatter; or a strong man pushing a defective car off the road when its engine has failed. And it is true that the concept of pushing and pulling is deeply woven into our notion of causation. However, in real life, and especially when we are seeking to ascribe responsibility for an event to a human, as opposed to a mechanical or natural agency, we are perfectly accustomed to regarding failures to act as possible causes. As J. S. Mill argued in his *Logic*, Book I Chapter v, if the watchmen fall asleep at their posts, and fail to keep guard, so that the enemy can enter the city, their failure is deemed to be the cause of the defeat. If I were a surgeon and for some reason refused to operate on an otherwise healthy person who had acute appendicitis, and he died, I should have caused his death through doing nothing just as certainly as if I had stabbed him. To decide not to operate on Baby Doe was to cause his death; and if killing a living human being is always wrong, then it was wrong to fail to operate. Incidentally, because of the prevailing view that it is one thing to withhold treatment, or withdraw it, and another to give a lethal injection; between allowing to die, that is, and killing, Baby Doe, though condemned to die, had to die by slow and long-drawn out stages of starvation and dehydration. It would undoubtedly have been more merciful to kill him outright once the decision was taken. But this is by the way. We have still to consider whether or not the killing was, in this case, justified.

The most familiar argument against allowing a baby to be killed in this kind of case is as a matter of fact the so-called 'slippery slope' argument. This argument has an astonishing power over the public imagination, but it must be distinguished from any argument based on the equal value of all human life, the principle that was incorporated, as we have seen, in President Reagan's memorandum. The slippery slope argument says that if the killing of babies like Baby Doe were permitted, then, even if that particular killing were justifiable, anything would become permissible. The next thing would be that parents could ask that any baby might be killed if they didn't like the look of him, or if they had decided that after all they'd prefer not to have a baby. And from babies we could move on to other children and grown-ups. We could kill them if they were handicapped, and then, further on down the slippery slope, we could kill them if they were Jews. The only way, so it is argued, to avoid hurtling down this morally disastrous slope is never to get onto it in the first place. No killing at all should be legitimised.

Against this kind of argument I would urge two considerations. First, it is perfectly possible to make a firm decision as to how far down the slope it is thought morally proper to go, and then to place a barrier, if necessary by means of legislation, beyond which further descent is to be deemed an offence. Thus, for example, The Committee of Inquiry into Human Infertility recommended that research using human embryos should be permitted only up to the fourteenth day from fertilisation, and from then on to use an embryo should be a criminal offence. I believe that, if implemented, this barrier would prove effective. Analogous regulations operate with regard to the use of animals in laboratories for research purposes. On the whole we may be confident that people will go no further down a slope than society wants them to.

Secondly, those who deploy the Slippery Slope argument tend to compare things at the top of the slope, which are agreed to be undertaken from a respectable motive, with things supposedly at the bottom, whose motive would be entirely different. Thus scientists wishing to use the pre-embryo for research are likened to Nazis using Jews for research before destroying them in gas chambers. A moment's thought will show the radical difference in motivation between these two kinds of activity, which are therefore not, as it were, at different ends of a single scale.

Let us then abandon the slippery slope argument and go back to the real issue, which is whether we are entitled to value human life non-uniformly giving more value to some lives than to others. If so, this must either be on the basis of the nature (or perhaps the age) of the human being who is alive, or on the basis of quality of his life, regardless of age. Sometimes considerations of age (or stage of development) may be hard to distinguish from considerations of quality. Obviously doctors, and lay people who have duties of care towards another human, are faced with these kinds of judgment of the quality of life simply because technology now allows us, if we wish, to keep alive, by means of life-support machines, patients who even a few years ago would certainly have died. Are doctors (as President Reagan seemed to assume) really bound to keep their patients alive as long as it is technically possible to do so, regardless of the kind of life the patient is now leading and will be leading in the future? The Hippocratic oath, for what it is worth, commits doctors to doing good and not harm to their patients. Most doctors accept this, but do not accept that being kept alive is *necessarily* a good for a patient. Will the patient live a life that is worth living, a life that we would value, or more importantly, that he would?

It is no longer possible, as once it was, for doctors to make their decisions secretly, perhaps in unspoken collaboration with nurse or midwife. For the public is now far more alert to what doctors are doing, and far more aware of what measures to prolong life are technically possible. Moreover, as in the Baby Doe case, there is a considerable number of people not personally connected with the patient but committed in principle to the prolongation of life in all circumstances, the anti-abortion, Pro-Life lobby who will be watching a doctor or nurse, ready to denounce him and bring him to court if he might be responsible for a patient's death, even though the relatives may be in agreement with him. And so we cannot go back to the old days; decisions must now be made publicly.

In this country, severe malformation of the foetus is a recognised ground for abortion, if the mother wants it (the right if any of the father to take part in the decision being still legally uncertain). Though there are hardliners who argue that even when a foetus is diagnosed as being an encephalic, or having only fragments of a brain, the mother should be obliged to continue the pregnancy to term, this is not the law, nor does it accord with the moral principles of most people, whether medical or lay. It seems to me therefore that at least in the case of neonates we ought to rethink legislation in such a way that doctors, in collaboration with parents where possible, are quite unambiguously entitled to end the life of those whose life-chances are seen to be very poor. As I have suggested the most commonly used argument against this is generally the Slippery Slope argument, and this we have put aside. Poor life-chances should include both the near-certainty of early death without the support of artificial means to continue breathing or heart-beat; but also the certainty of a life which, if prolonged, would contain an obvious preponderance of misery over satisfaction. We have here entered the thorny field of deciding how greatly we value the life of an infant by reference to the quality of that life.

It is obviously extremely difficult to assess quality of life on some-one else's behalf, yet if we are not committed to the pretence that we value the lives of all humans equally, simply because they are human, this is what we are bound to do. For it would be manifestly wrong to say simply that lives are less valuable as the person whose life it is gets older, or if they are unusually young (i.e. premature). Sometimes length of life may be relevant, but there can be no gen-eral rule. This being so, we are bound to consider *intrinsic* quality as the basis for value. The difficulty of judging the quality of other people's lives is especially hard when we are considering a living creature who will never reach a stage of self-awareness, or aware-ness of others, such as to enable him to compare his own lot with theirs. There is no doubt that a very severely mentally retarded person, unable ever to sit up or communicate or make any choices or discriminations in his surroundings, has a life almost incred-ibly poor compared with that of the rest of us. But does he suffer? Presumably not, provided that he is not actually cold, hungry or in physical pain. We would not like his life, any more than we would like the life of a mouse or a flat-worm, but of course to say that is to suppose that, while living one of those lives, we retained our own sensibilities. If we had mouse-sensibilities we'd do all right as mice. On the other hand we *can* judge the kind of disability which involves gradual degeneration with increasingly long spells of hospitalisation, where the child not only suffers pain and sickness but can see himself gradually changing, drawing away from his peers, and facing death. This is a kind of suffering that must be real and terrible for the child himself.

But it leads us to a second complication. Foreseeing this kind of life for their child, his parents may well choose, if they can, that the infant should die as an infant, knowing nothing. They value his life, and therefore the decision is an extraordinarily painful one; but they do not value all kinds of life equally, and so they may decide that he should die. In making this decision, are they to be allowed to consider their own future suffering, and perhaps that of other members of their family? Is it the future quality only of the infant's own life that is to be taken into account? I believe that parents are not to be blamed, still less prosecuted, if, faced with the prospect of severe mental retardation or non-function, or with that of progressive physical degeneration leading to early death for an infant, they prefer immediate death for that infant, partly on the ground that his living would be an intolerable burden to those who are already alive (or perhaps yet to be conceived). Our primary duty must be to those humans who have been born and who have embarked on a life which they can direct and evaluate; and here I believe that the role of the doctor is of tremendous importance. For parents may be quite unable to take the decision to allow a severely disabled child with poor life chances to die, because to do so is the equivalent to killing the child, although that would be in the same moral tradition as our concept of mercy. Is it not in the best spirit of that tradition that we temper the one with the other? We can embody these ideals best by not insisting that physicians invariably perpetuate lives that are painful and dolorous and by allowing parents the freedom to decide what is not an acceptable life for their child, using the human heart as their guide. This is a suggestion I think we have seriously to consider.

I will turn now to a different kind of case from that of Baby Doe and other neonates. The Committee of Enquiry into Human Fertility and Embryology, whose report was published in 1984 recommended that the human embryo, fertilised *in vitro* might be used for research if it were not needed for placement in the uterus, though only up to fourteen days after fertilisation. I need not go into the arguments that led us to place a cut-off point at fourteen days. These arguments are fairly well known and have been accepted by the Royal College of Obstetricians and Gynaecologists and the MRC whose voluntary guide-lines lay down such a limitation on research. In the present context the question is not where the limit should be placed, but whether *any* research using human embryos should take place. For there is no dispute about the fact that the embryos in question are both human and alive; so if all human life is of equal value, and if it is the case that humans should not be used in research programmes without their prior consent, though other animals may be so used, then research using human embryos should be criminalised just as certainly as research on other humans, without consent (and of course embryos cannot give consent). Moreover the fate of the embryos after they have been used for research is to be killed, and we have seen that the great difference between the claims of human beings and other animals is that humans must not be killed to meet the needs of other humans, while other animals may be, as well as being used in other ways for ends not their own. Does the fact that a creature is embryonic diminish its status as a live human? Those who are opposed to the use of live embryos for research would reply that it does not (and of course such people must, if they are to be consistent, also be opposed to all abortion). If human life is supremely valuable, they would argue, then it must all be equally valuable, and the age of the human creature, and the question whether or not it has been born, makes no difference.

Let us suppose that it is granted that research using embryos and fertilised ova is undeniably advantageous, not to those embryos and fertilised eggs but to other humans, including many yet unborn. The question then becomes this: valuing human life as we do, are we entitled to value the quality of future human lives more than we value present embryonic human life?

A utilitarian argument would suggest that we have no need to worry about the use of human embryos, provided that the outcome of the research will be a diminution of pain and an increase in satisfaction in the future. For embryos, at least before 14 days from fertilisation, are incapable of suffering, indeed, having as yet not even the beginnings of a central nervous system, are incapable of experiencing anything whatever, agreeable or disagreeable. Therefore they need not come into the calculus of pleasures and pains at all. However, having rejected utilitarian arguments in the matter of distinguishing human from other animals, we cannot reintroduce them at this stage, to justify the use of human embryos in research. Nor do I believe that we need to. We can instead consider, not so much the quality of the life of the embryo, as the nature of the structure of the embryo at this early stage of development, and derive from that a distinction between the value we should attach to it and its life and the value we do attach to a fully formed human and *its* life. I have already suggested, as a general principle, that we do, and should, value a human who has been born and is living his life, over a foetus, or even a severely damaged neonate who will never live a life fully his own. This bias in favour of the born over the unborn is still stronger when we consider the embryo a few hours or days after fertilisation. For one thing there is in nature a great wastage of embryos at this period of their development, many spontaneous abortions occurring without anyone's knowing it. Nature is prodigal with such embryos, and we cannot mourn them, as we may mourn a foetus in a later miscarriage or a stillborn baby. But more important, when we think of the value we attach to a human life, it is of the life of an irreplaceable individual that we think. I shall say something more about this at the end of my lecture. For the moment I simply want to *assert* that it is human individuality that we value. We strongly believe that human beings are not replaceable. One cannot substitute for another. If I am anxious about an endangered species of animal, I do not care which seal dies, or which seals are born. If I catch someone stealing the eggs of a peregrine falcon, I regret that there will be fewer birds, not that certain specific birds have been destroyed. It makes no difference *which* falcons survive as long as some do. We cannot adopt this attitude towards human beings. But towards the early embryo our attitude is necessarily different. It is impossible to think of the fertilised embryo, whether *in vitro*, or *in utero* as a specific individual. Although its genetic composition is there, it is still unclear whether there will, so to speak, be one of it or two, since twinning may still occur up to fourteen days from fertilisation. Nor is it certain which of the cluster of cells will form the embryo if one or two form at all, which will form the placenta. Though what we are dealing with is as it were human material, it is not yet a human individual for the cells are not yet differentiated.

Therefore the unique value that we attach to an individual human cannot be attached to the embryo, (or pre-embryo, as it is often called). We may therefore treat human embryos at this stage as indeterminate beings, and we should be able to use them provided that we are certain that their use will benefit others, either other specific humans, or other humans in general. Embryos, at the stage we are considering, are more akin to human tissue than to human children. Their lives are less valuable than the lives of *any* individual human being.

One of the uses of research using human embryos has been to allow the development of the techniques, first of gene identification, and then of possible gene therapy. This is an area of increasing importance and of increasing world-wide anxiety. Nowhere is the Slippery Slope argument more fervently and passionately deployed than in this area. For a nightmare future is foreseen in which eugenics will be compulsory; the ideal human will be sought and created; where whole races or classes of people will be eliminable by political edict. And, added to that, we are assured that the world will be full of vast animals, enormous birds and viruses with wholly unknown properties. Nowhere, therefore, is regulation more necessary, in order effectively to block off the dread slope, and this both in agriculture and human medicine. In fact the regulation has been in place for a long time in the context of agricultural research. It is in the medical field that we need now to seek for guide-lines, if necessary backed up by legislation; and such guidelines are being actively sought in America, Australia and Europe. Confining ourselves then to human medicine, there is a distinction to be drawn between 'germ cell therapy' and 'somatic cell therapy'. In the former, genetic changes would be brought about which would affect, not just one individual, but all of the offspring of that individual. In the latter, therapy is undertaken, the replacement of a single defective gene, which will affect only the individual in whom the defective gene has been identified. There is, I believe, general agreement at the present time that germ cell therapy should not be undertaken, since the outcome is too long-term to be predictable. Somatic cell therapy, on the other hand, can be seen as little different in principle from organ transplant, but on a molecular scale. Nevertheless it is plain that even in the case of somatic cell therapy, there is still a fear that the technique may in future be developed and used so as to change people, recreate them in a new image, manipulate them for political ends. Thus regulation must be introduced, and publicly justified.

Such regulation must take the form of the drawing up of a list of those areas of gene therapy which should, and those which should not, be actively pursued in research and practice. And it is in the drawing up of such lists that *value* judgments will inevitably play an important part. Let us, for the sake of simplicity, suppose (what is far from the case) that *any* gene can be identified, and *any* replaced. Which genes will merit treatment? One can imagine a short list of conditions which it would be widely agreed were intolerable, the elimination of which in individual cases would be totally acceptable. For example, Russell Scott, the Australian lawyer, though he agreed that gene therapy should be regarded as experimental, and should in any case be confined to the identification and replacement of somatic genes, wrote "the purpose of research in relation to gene therapy, and its justification, is the reduction of disease and human suffering. The potential benefits offered are therefore self-evident". He quoted the Gene Therapy guidelines, issued by the Australian Health and Medical Research Council in November 1987: "The choice of diseases ... is critical. Initial trials should be limited to diseases which cause a severe burden of suffering and for which there is no effective treatment". (NH and MRC Report on Round Table Conference on Human Gene Therapy, Canberra 1988).

It will be obvious from the very idea of such limitations and restrictions that we are quite prepared to form judgments about human life; at least we are sure that there are some genetically inherited disorders (like short-sight for example) which we believe have virtually no bearing on the quality of a person's life, even though they are imperfections, and others which we believe make life not worth living, so great is the "burden of suffering". We are confident, that is to say, about value-judgments which concern conditions at the extreme ends of a continuum of human disability. There are doubtless many other ethical problems that will arise as gene identification becomes more widely possible, problems connected with confidentiality and insurance and how we are to rate the discovery that someone is *liable* to, say, heart disease or schizophrenia, though not *certain* to suffer. But as far as the urgent regulation of medical research goes (and this is my concern) we can conceive of a system of regulation that will be based on an *evaluation of life-chances*, an evaluation of the *quality* of life, which will be very generally agreed.

I conclude therefore that we can and do distinguish, for other people as well as for ourselves, good quality of life from bad, bad from intolerable. And I have argued that, even though we are bound to value the lives of humans above the lives of other animals, we are *not* bound to value all human life equally, but we may distinguish the value of human lives, roughly speaking, in accordance with their quality. This entails our learning to draw an all-important distinction, and one which lies behind all our thinking about the value of life. This is the distinction between life in the biological sense and life in the sense in which we, members of the human species, have lives that we lead. A member of the human species *has* his experiences in the sense that they belong to him as an individual who identifies himself, who distinguishes his past from the past, his future, including his plans and hopes from the future. His life is, or will become for him, a story with a beginning, and a middle, if not yet with an end. He is the centre, the hero of the plot. I could not lead my life if I were not alive, but it is *my* life that I value, and out of sympathy and moral duty I value the lives of others, because I know that for them too there is a plot, a drama, of which they, each of them, is the hero. The different stories overlap and interlock, but none is identical with any other. The life of a newborn baby is valuable because his life is just beginning to unfold for him, and we know that he will become aware of it and start to shape it. But if there is a baby of whom we know that he never will do this, either because he will die too soon, or because he will never reach that state of cognitive ability in which he will be aware of himself *as* himself, then we must value his life less, though he is alive. Similarly if someone has lived his life and is now in an irreversible coma, though still alive, it is idle to pretend that we now value his life as highly as we did before, provided there is no hope of his recovery. Nor should we pretend that we ought to do so.

I believe that it is of the greatest importance to draw this distinction. For if we acknowledge that it is not simply *human life* that we value, all equally, but *individual lives to be led*, then we can form our conclusions on a whole number of ethical problems, including that of abortion, our treatment of severely disabled neonates, research using human embryos, and, in the future gene therapy. Attempting to solve these problems will no longer seem a matter of metaphysics, but of legal, medical and scientific judgment.

Appendix:

A List of the Eddington Memorial Trust Lectures

Nos. 1–26 and 28 were published by Cambridge University Press.
No. 29 was published by the author in *Contemporary Physics*, Vol. 29 (343–405), 1988.
No. 30 is believed to have been published by Mary Warnock in a collection of her talks and speeches.
Nos. 27 and 31–34 were not published.

NO.	TITLE	LECTURE DATE	AUTHOR
1	Reflections on the Philosophy of Sir Arthur Stanley Eddington	4 November 1947	A. D. Ritchie and C. E. Raven
2	Sir Arthur Stanley Eddington – Man of Science & Mystic	12 November 1948	L. P. Jacks
3	The Unity of Knowledge – Reflections on the Universities of Cambridge and London	1 November 1949	George B. Jeffrey
4	Creative Aspects of Natural Law	2 November 1950	R. A. Fisher
5	Eddington's Principle in the Philosophy of Science	9 August 1951	Sir Edmund Whittaker
6	Time and Universe for Scientific Conscience	4 November 1952	Martin Johnson
7	Some Aspects of the Conflict between Science & Religion	3 November 1953	H. H. Price
8	The Sources of Eddington's Philosophy	2 November 1954	Herbert Dingle
9	An Empiricist's View of the Nature of Religious Belief	22 November 1955	Richard B. Braithwaite
10	Thought, Life and Time, as Reflected in Science & Poetry	19 February 1957	H. G. Wood
11	Science and the Idea of God	21 April 1958	Charles A. Coulson
12	Science, Philosophy & Religion	22 April 1959	Russell Brain
13	Beyond Nihilism	16 February 1960	Michael Polanyi
14	Biology, Psychology & Belief	3 November 1960	William H. Thorpe

15	The Vision of Nature	10 November 1961	Sir Cyril Hinshelwood
16	On Having A Mind	9 November 1962	William Kneale
17	Mind & Consciousness in Experimental Psychology	29 November 1963	R. H. Thouless
18	I Believe…	6 November 1964	Kathleen Lonsdale
19	The Brain and the Unity of Conscious Experience	15 October 1965	John C. Eccles
20	Scientific Principles & Moral Conduct	15 November 1966	James B. Conant
21	Freedom of Action in a Mechanistic Universe	17 November 1967	Donald M. Mackay
22	Time, Change & Contradiction	1 November 1968	G. H. von Wright
23	Observations on Man, His Frame, His Duty and His Expectations	7 November 1969	W. Grey Walter
24	Abstraction in Science & Morals	2 February 1971	S. Koerner
25	Biology and the Soul	1 February 1972	J. H. Hick
26	The Challenge of the Third World (series of three)	7, 14, 21 November 1974	Joseph Hutchinson
27	Science and the Divine: an historical approach to the problems of science and the world today (series of four)	May 1978	Maurice Wilkins
28	From Paracelsus to Newton (series of three)	October 1980	Charles Webster
29	The Invincible Ignorance of Science: the mind–matter chasm	28 January 1988	A. Brian Pippard

NO.	TITLE	LECTURE DATE	AUTHOR
30	The Value of Life	26 January 1989	Mary Warnock
31	Objectivity and the Subject: scientific understanding and subjective perceptions of ill-health, poverty and women's inequalities	May 1990	Amartya Sen
32	The Emperor's New Mind	February 1991	Roger Penrose
33	Science & Imagination	30 October 1992	Francois Jacob
34	Living Molecules	24 November 1994	Max Perutz
35	Law and the scarcity of medical resources	7 May 1999	Len Hoffman

Printed in the United States
by Baker & Taylor Publisher Services